Identity, Aesthetics, and Sound in the Fin de Siècle

This book is an analytic and historical portrait of the volatile decades at the beginning of the 20th century. Engaging with avant-garde art and thought, and concentrating on two of the most controversial and still culturally relevant personalities of Viennese modernism—Sigmund Freud and Arnold Schoenberg—it tells the story of a cultural experiment of unprecedented proportions, an experiment that attempted to redesign the senses and the concept of individual identity. The book describes the shape of this identity through its mutually overlapping artistic and intellectual dimensions, as it explores the relationship between psychoanalysis and music.

Dariusz Gafijczuk is currently teaching and researching at the School of Slavonic and East European Studies, University College London.

Routledge Studies in Cultural History

1 **The Politics of Information in Early Modern Europe**
Edited by Brendan Dooley and Sabrina Baron

2 **The Insanity of Place/ The Place of Insanity**
Essays on the History of Psychiatry
Andrew Scull

3 **Film, History, and Cultural Citizenship**
Sites of Production
Edited by Tina Mai Chen and David S. Churchill

4 **Genre and Cinema**
Ireland and Transnationalism
Edited by Brian McIlroy

5 **Histories of Postmodernism**
Edited by Mark Bevir, Jill Hargis, and Sara Rushing

6 **Africa after Modernism**
Transitions in Literature, Media, and Philosophy
Michael Janis

7 **Rethinking Race, Politics, and Poetics**
C. L. R. James' Critique of Modernity
Brett St Louis

8 **Making British Culture**
English Readers and the Scottish Enlightenment, 1740–1830
David Allan

9 **Empires and Boundaries**
Rethinking Race, Class, and Gender in Colonial Settings
Edited by Harald Fischer-Tiné and Susanne Gehrmann

10 **Tobacco in Russian History and Culture**
From the Seventeenth Century to the Present
Edited by Matthew P. Romaniello and Tricia Starks

11 **History of Islam in German Thought**
From Leibniz to Nietzsche
Ian Almond

12 **Israeli-Palestinian Conflict in the Francophone World**
Edited by Nathalie Debrauwere-Miller

13 **History of Participatory Media**
Politics and Publics, 1750–2000
Edited by Anders Ekström Solveig Jülich, Frans Lundgren and Per Wisselgren

14 **Living in the City**
Urban Institutions in the Low Countries, 1200–2010
Leo Lucassen and Wim Willems

15 **Historical Disasters in Context**
Science, Religion, and Politics
Edited by Andrea Janku, Gerrit J. Schenk and Franz Mauelshagen

16 **Migration, Ethnicity, and Mental Health**
International Perspectives, 1840–2010
Edited by Angela McCarthy and Catharine Coleborne

17 **Politics of Memory**
Making Slavery Visible in the Public Space
Edited by Ana Lucia Araujo

18 **Neutrality in Twentieth-Century Europe**
Intersections of Science, Culture, and Politics after the First World War
Edited by Rebecka Lettevall, Geert Somsen, and Sven Widmalm

19 **Americans Experience Russia**
Encountering the Enigma, 1917 to the Present
Edited by Choi Chatterjee and Beth Holmgren

20 **A Social History of Disability in the Middle Ages**
Cultural Considerations of Physical Impairment
Irina Metzler

21 **Race, Science, and the Nation**
Reconstructing the Ancient Past in Britain, France and Germany
Chris Manias

22 **Identity, Aesthetics, and Sound in the Fin de Siècle**
Redesigning Perception
Dariusz Gafijczuk

Identity, Aesthetics, and Sound in the Fin de Siècle

Redesigning Perception

Dariusz Gafijczuk

Routledge
Taylor & Francis Group
NEW YORK LONDON

First published 2014
by Routledge
711 Third Avenue, New York, NY 10017

Simultaneously published in the UK
by Routledge
2 Park Square, Milton Park, Abingdon, Oxon OX14 4RN

Routledge is an imprint of the Taylor & Francis Group, an informa business

Library of Congress Cataloging-in-Publication Data
Gafijczuk, Dariusz.
Identity, aesthetics, and sound in the fin de siècle: redesigning perception / by Dariusz Gafijczuk.
pages cm. — (Routledge studies in cultural history ; 22)
Includes bibliographical references and index.
1. Vienna (Austria)—Intellectual life—20th century. 2. Freud, Sigmund, 1856–1939—Philosophy. 3. Schoenberg, Arnold, 1874–1951—Philosophy. 4. Psychoanalysis and music—Austria—Vienna—History—20th century. 5. Avant-garde (Aesthetics)—Austria—Vienna—History—20th century. 6. Senses and sensation—Austria—Vienna—History—20th century. 7. Identity (Psychology)—Austria—Vienna—History—20th century. 8. Perception—Social aspects—Austria—Vienna—History—20th century. 9. Sound—Psychological aspects—History—20th century. 10. Hearing—Psychological aspects—History—20th century. I. Title.
DB851.G25 2014
943.6'130441—dc23
2013011439

ISBN13: 978-0-415-70428-1 (hbk)
ISBN13: 978-0-203-76214-1 (ebk)

Typeset in Sabon
by IBT Global.

SUSTAINABLE FORESTRY INITIATIVE Certified Sourcing www.sfiprogram.org SFI-01234
SFI label applies to the text stock

Printed and bound in the United States of America
by IBT Global.

To Maia, who is already redesigning my perceptions

Contents

Introduction

> The forms of art register the history of humanity with more justice than do historical documents.
>
> —Theodor Adorno

> Every civilization, both yesterday and today, is immediately manifest in something easily grasped; a play, an exhibition of paintings, a successful book, a philosophy, a fashion in dress, a scientific discovery, a technological advance—all of them apparently independent of one another.
>
> —Fernand Braudel

This book is an analytical portrait and a historical reconstruction of cultural perception at the beginning of the twentieth century. My engagement with this theme focuses on the avant-garde thought, broadly defined. But a double and somewhat unusual inflection runs a gamut of reincarnations throughout the text, embodied in the intellectual and artistic profiles of two figures seemingly on the opposite ends of the creative process—the founder of psychoanalysis, Sigmund Freud, and the composer Arnold Schoenberg. Apart from the shared context of Vienna circa 1900, could there be a more meaningful and deeper connection between them? Working with the contextual and conceptual evidence that points to such a deeper affinity, the following argument recovers a series of overlapping dimensions from the context Freud and Schoenberg inhabited—an environment that since then has become the most popular stand-in for all things modern.

The many differences between psychology and music notwithstanding, there are indeed numerous points of convergence between sound and psyche at the beginning of the twentieth century—some of them surprising and quite a few unexplored. It is my contention that these convergences allow us to glimpse the years around 1900, usually characterized as the time of directionless confusion brought about by modernity reaching its 'psychological apex', in a different profile—as a project working restlessly to redesign cultural perception from the ground up and coming remarkably close to doing so on an unprecedented scale. My encounter with the *fin de siècle*, a formulation that, by all appearances, was launched in 1888 by F. de Jouvenot and H. Micard in a play of the same title, tells the story of how by 1900 'end-of-the-century' became much more than just a temporal incision into the flow of time, but a cultural experiment of immense proportions—an experiment that worked according to the blueprints of expressive *transposition* between various spaces and cultural registers. Cultural perception, sensory configuration, and the identities based on them were continually shifted into adjacent spaces—some of them closely related and others surprisingly distant. This experimental space was based on the dynamics of resonance and the ever extended range of various 'amplification technologies', of which psychoanalysis is perhaps the best-known example.

This is how the conditions for Central Europe (because this is where we will remain most of the time in our narrative) to become "the laboratory of twilight", as Milan Kundera refers to it—a type of "premonitory mirror showing the possible fate of all of Europe"[1]—were created. That fate would eventually lead to a complete undermining of the reassuring symmetries connecting the modern way of life with the growing anonymity of the world. This book describes the nascent shape of such a malleable identity, recovering some of its parameters from the much broader field of artistic, conceptual, and cultural topographies forged a mere hundred years ago.

The decades at the beginning of the twentieth century hold a firm strategic outpost in the contemporary cultural imagination. These vertigo years, as Philip Blom has recently described them,[2] full of unforeseen and more crucially unimaginable consequences, are the closest we have come to something like the founding mythology of the present time. In other words, the decades surrounding the year 1900 fused together fact with fiction, myth with science, art with politics. Each and every one of these overlapping dimensions, in its special way, carried the condensed logic of the entire cultural and political setting within its apparently restricted field of expression. Thus, as mentioned already , I framed this book around two of these—at first glance distant and more importantly irreconcilable phenomena forged in the cultural laboratory circa 1900—Sigmund Freud's psychoanalysis and Arnold Schoenberg's acoustic philosophy, because in the end, Schoenberg does much more than simply compose music.

This intersection of what on the surface may appear simply as an intellectual and an aesthetic phenomenon is an example of the type of tension and volatility that informed the entire cultural landscape at the time. Such a tension, usually invisible to the 'naked' historical eye, must be recreated, bringing us as close as possible to its many subtleties, paradoxes, and hidden potentials. My venture into the beginning of the last century does this by juxtaposing perception and acoustics in ways that traditionally have been overlooked. But a series of such connections, of which Freud's 'talking cure' and Schoenberg's sound based on the constantly redesigned relationship between musical tones are arguably the starkest, was generated by the historical fabric of the *fin de siècle*, and we will have to see how. The most complicated aspect of this overlap between psychology and resonance underpinning artistic and intellectual production at the time is the realization that such 'acoustic logic of perception' was not merely embodied in sound but embedded in the entire spectrum of communication, including vision.

At its most comprehensive, the *fin-de-siècle* project was an attempted retuning of modern, psychological sensitivity. As such, new media of dissemination were also needed to effectively communicate its message. The most innovative of these was a new type of aesthetics based on the mechanics of resonance and its inherent plasticity. Its reach was felt across the cultural continuum. There were places resonance was primary but not exactly what one would expect, such as Freud's psychoanalysis or Arnold Schoenberg's 'music', and spaces where it was fundamental but where officially it was not supposed to be found at all, such as expressionist painting or the architecture of Adolf Loos. What happened in between these positions was a controlled descent into the frenzy of cultural self-expression.

The story of such a descent belongs to the cultural history of *aural* subjectivity that has always run parallel to the representation of the intellect based on the surveying power of the eye.[3] The three centuries before the *fin de siècle*, on balance, attempted to define modern individuality through an increasingly coherent relationship between the eye and the ear,[4] but around 1900 that project collapsed under its own weight. Suddenly, the task was to position oneself in the cavity left by the erosion of grand narratives, and any attempted rescue of the self-assured, centralized reality, including the concept of the individual, had to find ways towards its productive dissipation,[5] and this in turn meant losing the classical stability of carefully measured perspectives.

The origins of an identity based on the dynamics of disappearance can be traced back to at least the eighteenth century. This is when the project of crafting new, enlightened subjects also set the limits of perception adrift, adding to its creative repertoire whatever could extend it, including sound, and in the process creating the conditions for what at the turn of the twentieth century, especially in Freud's hands, would become known as depth-psychology.[6] Depth-psychology cannot function without a special type of plasticity based on resonance. This is how the process of 'wearing away' (abreaction), so fundamental to the psychoanalytic technique, is triggered.

***

The cultural experiment in redesigning perception , in all its facets, was especially intense in the region of Central Europe, with Vienna as the 'catch-all' signifier. This "little world in which the big one holds it tryouts",[7] as the often varied and repeated phrase suggests, was not simply an epiphenomenon of a series of failures, especially the failure of liberal politics,[8] but a collective version of what Freud described as *reality testing*. Failures of all kinds there surely were, but these equally constituted so many consequences of the 'natural' erosion or political manipulation of the traditional cultural fabric, as they were the result of a deliberate attempt to redesign the totality of lived space, in its various dimensions. This special brand of *fin-de-siècle* consciousness seems to have been forged in the increasingly stressed textures of time itself in which various intellectual and cultural phenomena found the right proportion of what Goethe described as 'elective affinity' (*Wahlverwandschaft*)—sheltered in the volatility of the many complexes, mixed up and deranged perceptions, intellectual formulas, and aesthetic forms, for which, the turn of the century is so famous . And the most challenging question then and now is—what connected all these phenomena?

The first of these universal connectors was the unprecedented encounter of everyday life with culture as an event in its own right. This meant that culture changed from something that was inherited and learned through a long and painstaking process of what was known as *Bildung* (education/cultivation) to something that could be created quickly and, most importantly, manipulated at each and every moment, as the mode of expression dictated. The result was that individual identity and collective consciousness became more analytically isolated than ever before. We need only think of one instance here in relation to Vienna 1900—the Secession movement led by Gustav Klimt, and its clearly stated aim to strip reality bare by searching for the 'naked truth' behind mere appearances. A newly found temporal consciousness seems to have been a crucial force here, as is clearly surmised by reading the famous motto etched into the stone façade of Josef Olbrich's 1898, purpose-built Secession house—*Der Zeit Ihre Kunst/Der Kunst Ihre Freihet* ("To the Times Its Art/To Art Its Freedom). This is in clear contrast, for instance, to Schorske's point of contact with Vienna 1900 as "a-historical" culture, where relationships are formed and defined through their independence of the past—something that, seemingly , also strikes at the heart of the modern condition.[9] However, as has become increasingly obvious, modernity is not so much indifferent to time or history, as it is a new form of awareness that redefines the parameters of both, bringing the past and the present together closer than ever before. Psychoanalysis, for instance, is literally unthinkable without such a temporal consciousness, although it is a strange, inverted sort of time—a vertical duration that allows Freud access to the seemingly immemorial past.

At stake in this search for the new times was a total recomposition of personality and through it a reconstruction of the human form—its mode of expression and the accompanying psychology. In other words, the idiosyncratic character of this particular moment in the history of modernity, with Vienna 1900 as its 'banner identity', allowed a new type of space to emerge—a space based on the model of productive dispersion of things and identities because, according to Ernst Mach, both were not salvageable in their current form given how even the "ego is as little absolutely permanent as are bodies."[10]

Such historically generated discourse of impermanence also applies to the urban topography of Vienna—its material and metaphorical extensions. This is the reason why I have decided not to accord the context of Vienna as such the distinction of a special case—a type of self-propelling reality elevated to the position of uncontested leadership in the history of recent culture.[11] Its importance notwithstanding, Vienna was just one nodal point in the wider avant-garde modernity, and certainly not the most significant one in all cultural and artistic dimensions.[12] Rather, the context of Vienna will perform a function similar to a screen memory, or a point of contact for imagination, which, according to Freud, meets us halfway and frames the simultaneous occurrence of certain events.[13] As such, my analytical attitude is akin to Schoenberg's disposition towards the city, expressed in a letter to Alma Mahler, when Gustav Mahler was on his deathbed: "I don't even feel that there is a Vienna anymore. For me it was always only a city in which this or that personality lived."[14] In this sense, Vienna stands as a somewhat confused participant in the 'age without qualities', to paraphrase Robert Musil's description of the years before World War I.[15] This is why I let Vienna as such disappear from most of the discussion—especially the Vienna of the Ringstrasse, coffee houses, the waltz, military parades, and afternoon strolls through the Prater. This is how a point of departure presented itself not in the shape of a specific place and its identification as something already and extensively spoken about but in the form of a somewhat hapless setting and its hidden potential. Such a potential, and the special kind of history that emerges from it, has always been snared halfway between sound and silence—a disposition that runs along the spine of our relatively recent past. Events and personalities, adrift at the beginning of the twentieth century, will be placed on its meandering trajectory marked by what Alain Badiou refers to as the "passion for the real."[16] This brings me to the second and more subtle of the elements constituting the shared logic that connected the phenomenal mêlée of events and spaces "paved with culture"[17]—sound and resonance.

Acoustics, until then largely confined to the safety margins of the musical paradigm, or used in highly specialized scientific fields, burst through its traditional boundaries. Ever more rapidly sonority came to the surface, recomposing the structure of various cultural dimensions, from painting and language to sculpture and even architecture.

We can look as far back as the mythological beginnings of 'Western' self-image to encounter this curious interplay between psychology and sound. It is thus all the more intriguing that the forgotten element in the saga of Narcissus's infatuation with his own reflection that at one point became one of the most identifiable monuments to modernity, is precisely sound.

Narcissus's misrecognition of himself was in fact a sentence pronounced for an auditory transgression—his refusal to hear Echo's invocations of love. This founding act of deafness repeats itself throughout the many representations of what we have come to know as modernity, leading to an obvious question: what would an identity not based on the narcissistic blindness but on the echo of incessant reverberations of our presence in the surrounding space 'look' like? The possibility of recuperating its prototype from a specific moment of cultural history is the task at hand, given that this is precisely the quandary one attempted to resolve in altogether unique form around 1900.

Attempting to find an answer to this precise question, Oskar Kokoschka decided to treat bodies "as the voice of the psyche."[18] It was the same voice without which psychoanalysis would have fallen into pure abstraction, not a practice, and Schoenberg's music would have lost its existential fervor, falling into the clutches of mere style or technique. The essential aim that motivated the artistic and intellectual field of expression was the search for as complete a paradigm of perception as possible. Schoenberg thought that no musical event, as a fully saturated phenomenon, can occur without at least an awareness of this perceptive dimension. In fact, throughout his creative life he searched for much more—a workable model of a unified theory of perception.[19] In 1917, as an already mature composer in full command of his craft, Schoenberg wrote,

> We consciously and unconsciously imitate in tones what we have perceived, even with different senses, in other materials. We also apply our way of thinking (of every kind) to the material of music.[20]

There is hardly a better formulation to describe the type of analysis carried out in the text that follows. Each chapter attempts to think with different materials and dimensions, while drawing analytical connections among various themes and fine-tuning our sense of 'the character profile' animating the years at the turn of the last century.

As such, I treat the *fin de siècle*'s deluge of conceptual formations as 'historical blisters' left on the skin of the present—full of encoded information about the cultural experience at the time. These are the elements that were used in the construction of the modern life as we know it today, including in its overdesigned and overmanaged forms. The years around 1900 ignited this kind of cultural impression management by carving out space for a new type of aesthetic psychology—doing so *en masse*. Such newly found orientation was no longer characterized by the traditional experience of distance as the transcendental overcoming of the here and now. Crucially, at one point, perception became trapped in the structure of cultural and later political *events* of various shapes and magnitudes. And we have perhaps not completely accounted for the profound change in cultural constitution when it is the *contingency* of events stripped bare that serves as the exclusive point of contact with the reality at large.

1 The Acoustic Symptom

The acoustic symptom has a history. In the version we know well it reaches back to around 1850 and, not unlike its mythological origins, it first deals with repression and then the recovery of echoes.

For what are symptoms if not echoes—acoustic fragments of ailments' invisible origins? In their physical and plastic amplitudes, they are transcripts of a mute totality—pieces of the whole, whose identity is constructed through a series of performances. In other words, symptoms are contractions reworking the overall context of an illness, tapering it down to a certain shape with a unique surface and texture. The important thing to note about such reconstructive logic is that it "does not describe a state of things, but immediately produces a real fact."[1]

Performance and the production of facts in symptomatic arrangements played a major role in the activities of Schoenberg and Freud. Their range extended from the more official and public instances of theoretical/practical elaborations (Freud's publication of case studies, Schoenberg's concerts), the semiprivate occurrences (relationships with students, patients, followers), to finally the very private and biographical. One such symptom that framed Freud's biography is his strained relationship to music proper—its early dismissal and misrecognition. This appears to be the case at least on the surface, because eventually the lining of sound will envelop the psychoanalytical habitat in its entirety.

The acoustic censorship began early. In his student days, Freud occupied the cramped family flat in Leopoldstadt, the traditional Jewish quarter of Vienna:

> This apartment, to which they [the Freuds] moved in 1875 . . . was scarcely lavish for the sizable family. Alexander, the youngest, Freud's five sisters, and their parents crowded into three bedrooms. Freud alone had his "cabinet" for his private domain, a room "long and narrow, with a window looking on the street," more and more crammed with books . . . If Freud's needs clashed with those of Anna or the others, his prevailed without question. When, intent on his school books, he complained about the noise that Anna's piano lessons were making, the

> piano vanished never to return. The Freuds must have been among the very few middle-class Central European families without a piano.[2]

Among these, there was another family that lived in the second district of Vienna at the time and shared a similar experience of silence: "It seems that there was no music-making in Schoenberg's parents' house. Even after the father had opened a collection agency the family remained in modest conditions and did not own a piano."[3] Of course, we should not overestimate such anecdotal evidence. Yet, taken at the level of the symptomatic, these circumstances speak to a certain collusion of circumstances that inform the auditory attitudes underpinning Freud's and Schoenberg's mature relationship to sound.

They start from two opposite ends. Freud progressively fills the silence that was so precious to him during his study years with his own instrument for sound making. This process begins in 1886 with the publication of a medical paper on the development of the acoustic nerve.[4] Schoenberg, who, given his early childhood experience, never acquired full instrumental proficiency, progressively extends this logic of detachment to his entire creative output, eventually dispensing with the standard acoustic identity, in a way silencing music to liberate sound. Both men will eventually meet in the middle, constructing systems of thought and practice that not only speak to each other but also contribute immeasurably to the character profile of the *fin de siècle.*

From the vantage point of our argument, the interesting aspect of these two parallel episodes is the early breach of standards in Freud's and Schoenberg's immediate environments. Its presence is felt through a range of symptoms. Of special importance is the acoustic interruption or the blockage of sonority, because sound in this case has been accused of luxury or noise, and declined. To a certain extent, this creates the side-effect condition of arrested development and censored performance of the bourgeois identity. Freud, through his objection, not only denies the pleasure of music to his sister (a pleasure that, as he will say some years later, he is unable to take part in)[5] but also inadvertently blocks the consumption and the expression of such a conventional identity to Anna and the family. This pattern will be repeated again and again by Freud himself, especially in his practice and the nexus of ties that bind the patient to a series of stammers and interruptions, returning him or her to a zone of subjective morphology that instigates a divergence from the previously accepted norm.

Freud's disavowal of sound acquires further symptomatic significance that goes to the heart of the psychoanalytic technique itself, for psychoanalysis is the performance of complaint and voicing of grievances through the production of sound. Given this, Freud is unable to escape sonority, leaving him with no choice but to develop an entire philosophy of 'soundings' and the 'technology' for their recording and interpretation. Psychoanalysis is indeed the treatment *with* and through sound, and its productive output is central to the technique that from the start is known as the 'talking cure', even if at first there seems to be a large gulf separating musical sound and speech, given how the latter is overburdened by meaning. However, a parallel version of grammatical oversignification is also endemic to music in the form of tonality and harmony as the syntax of composition, against which Schoenberg will stake a claim initially through a complaint, then an accusation, and finally an all-out attack. Therefore, on the most basic level and in various proportions, both Freud and Schoenberg manipulate what we may refer to as the *material-acoustic* phenomenon:

> The material element, which in all aesthetic enjoyment is at the root of the intellectual one, is greater in music than in any other art. Music, through its immateriality the most ethereal art, and yet the most sensuous one through its play of forms without any extraneous subject, exhibits in this mysterious fusion of two antagonistic principles a strong affinity for the nerves, those equally mysterious links in the invisible telegraphic connection between mind and body.[6]

Freud becomes the master of such telegraphy, at one point even comparing the role of the analyst to a telephone receiver—triggering the process of circular interpretations based on sequences of looping, interminable transmissions from the deepest, sonic frontiers of the psyche. Hermann von Helmholtz, who in his 1863 work *On the Sensations of Tone as the Physiological Basis for the Theory of Music* established the famous modeling of sound based on acoustic overtones—the internal echo of a note through which other, more distant tones are heard, leading to 'natural' harmonic affinities between them—also thought of the entire nervous system as networks of communication at a distance. "The nerve fibers have been often compared with telegraphic wires traversing a country"[7]—he wrote. New technologies have been frequently mapped onto the human body, unlocking its physical dimensions and transposing them onto new plateaus of perception. Something very similar took place between the psychology of the unconscious and resonance as the only tool that can penetrate deep enough to come into contact with its surface.

The unconscious can be delivered for interpretation only in packets of *aural information*. Such a setup operates on the basis of releasing the acoustic element from behind a soundproof partition—a mechanism surprisingly similar to the distinction between musical and pathological hearing introduced in the 1850s by Eduard Hanslick.

Perhaps the best-known music critic in Vienna from the mid-nineteenth century onwards, Hanslick made his name writing on the aesthetics of music. Careful reflection on the nature of sound led him to espouse an especially unusual position at the time: early and persistent criticism of Richard Wagner. In his best known work, *The Beautiful in Music* (first published in 1854, and reprinted nine times in its German edition) Hanslick argued against the still prevalent appraisal of music as an art, whose most basic and valuable merit is found in the affective arousal: "I firmly adhere to the conviction that all the customary appeals to our emotional faculty can never show the way to a single musical law" (*Beautiful in Music*, 4). It follows then that the inquiry into musical aesthetics is "mainly and primarily directed against the widely accepted doctrine that the office of music is to represent feelings" (*Ibid.*, 4). Feelings are extraneous notions that are motivated by the psychic/nervous state of the body, but strictly speaking they are not musical phenomena, because the latter are enclosed in their own form as sound: "The beauty of a composition is *specifically musical*, i.e., it inheres in the combinations of musical sounds and is independent of all alien, extramusical notions" (*Ibid.*, 5).

Emotion, presenting us with false assumptions about the 'nature' of sound, perpetuates the condition of *pathological hearing*, because it misidentifies sound in its most basic character that is always equal to itself. True hearing, on the other hand, is an intellectual and contemplative activity, a process of *attentive listening* to the sequences and combination of tones. The consequence of this state of affairs is that

> The word *Anschauung* (viewing, contemplating) is no longer applied to the visual processes only but also to the functions of the other senses. It is, in fact, eminently suited to describe the act of attentive hearing, which is nothing but a mental inspection of a succession of musical images.[8] (*Ibid.*, 11)

When, at one point,, Freud concludes that he cannot take pleasure in the art of music because one is not able to engage it through contemplation, he will practice his own brand of pathological hearing.

Even if the reality of feeling something while experiencing sound cannot be denied, there is no direct connection between music and emotions, because the latter do not originate in the acoustic structure but are imported into it by misdirected perception:

> For, in reality, there is no causal nexus between a musical composition and the feelings it may excite, as the latter vary with our experience and impressibility [sic]. (*Ibid.*, 14)

This is how, in the most intriguing manner, Hanslick discovers the blueprint for acoustics as a form of therapy that only a few years later will work itself into Freud's modeling of the psyche. The psychoanalytic patient is a case study in deformed perception caused by the pathology of hearing that is unable to receive the unconscious—a condition that Freud attempts to corner in its own distortion by instituting attentive listening as the proper approach towards the subject. He does so under the technical name of 'evenly suspended attention' (*gleichschwebende Aufmerksamkeit*). Even suspension of attention aims to eliminate the emotive factor by not unduly elevating certain kinds of content over others. This is why the acoustic-material phenomenon will replay itself and find shelter in Freud's psychological matrix, operating according to the free association of forms, not meanings as in language. If "the essence of music is sound in motion" (*Ibid.*, 48), then sound is the seat of unadulterated elastic materiality that works by always recombining and renewing the sequential elaboration of tones:

> The musical material in the hands of creative genius is as plastic and pliable as it is profuse . . . The union of sounds (from the interdependence of which the beautiful in music flows) is not effected by mechanically stringing them together but by acts of a free imagination. (*Ibid.*, 52)

The fundamental mechanics of plastic manipulation and elaboration can be characterized as a directed *interpretation* of sonority, a kind of *sound-work* (*Klangarbeit*). This is not only a pattern whose bravado will be explored by Schoenberg to its limit but also a similar type of procedure that will underpin Freud's analysis of dreams as dream-work (*Traumarbeit*). Dream-work and the technique of interpretation at its core depend on the pliability of forms—a principle that becomes the backbone of psychoanalysis.

Hanslick's model of musical aesthetics is a dynamic one. Sound is a phenomenon that moves and paces. And it may even suggest feelings, but it is not the source of these as such. Thus, it operates in a similar way to the Freudian psyche, collecting impulses into arrangements through condensation, displacement, and imaginary representations that step into the shoes of reality. Only these forms allow for an amplification of the deeper, independent psychological elements, but they never constitute their total replacement. Hence, Freud will never fully dispense with the dual model of the mind, even though he will come quite close to doing so. This is because the amplitude of psychological processes always remains suspended between their deeper composition and their rewiring for sound. Hanslick again anticipates the psychoanalytic session some fifty years prior to its invention, when he writes,

> A state of mind manifests itself most directly in music when *origination and execution coincide.* This occurs in the freest form of extempore playing, and if the player proceeds not so much according to the strict methods of art as with a predominantly subjective tendency (a pathological one, in a wider sense), the expression which he elicits from the keys may assume almost the vividness of speech. (*Ibid.*, 76, my emphasis)

In Freud's hands, the simultaneous presence of origination and execution will become known as the psychoanalytic session—the set for the directed performance of the analyst and the patient. It is a type of play that elicits forms channeled through the medium of sound out of the raw, 'unconscious', and chaotic material. Freud will later describe this type of engagement as *Deutung*, or 'interpretation', even though the German word is much more resonant than the abstractness of 'interpretation' has the ability to convey. In this sense, interpretation is more like sculpting in sound derived from subtle, textural contrasts, allowing Freud at one point to say that it is actually more appropriate to speak of psychoanalytic *constructions*, rather than interpretations.

***

If music is this ineffable, nondiscursive potency that pushes us along a trajectory of approaching the element that "directly, in itself . . . signifies nothing",[9] as Vladimir Jankélévitch says, how can we still cling to the idea of communicating with anything at all? This is indeed the core of the problem facing Freud. Jankélévitch, having the benefit of time as an ally, writing in the intellectual climate of postwar France, resolves this tension much in the same way Freud is inadvertently forced to do—by distinguishing between discursive and immediate, or situated communication, that in the manner of a performance produces immediate facts. This situates us in the "penumbra of melancholia" that can take place only in one direction, like "from hypnotist to hypnotized" (*Music and the Ineffable*, 9). This is where the psychological tension is harnessed to sound and released through the malleability of its form in the moment of reproduction, because music, like the psyche, "has broad shoulders. In the hermeneutics of music, everything is possible, the most fabulous ideologies and unfathomable imputed meanings" (*Ibid.*, 11).

Such transposition into a different register attempts to counteract the process of paralyzing overidentification with one, singular meaning. Psychoanalysis rebuffs it, replacing it with an entirely different circuitry of communication that organizes events acoustically, through segmental information presented concurrently along several planes. Unlike Narcissus, these are familiar with their echo:

> The melody's mirror-reflection of itself, and with canonic imitation . . . In polyphony, the voices speak together, harmoniously, but they are not speaking among themselves, *to one another*, they are not addressing themselves *to one another*: they are signing in concert for an outsider. (*Ibid.*, 20)

And it is in such a manner that they eventually sing to Freud.

Jankélévitch, who, as we now realize, says something quite similar to what Eduard Hanslick proposes, perpetuates a psychoanalytic understanding of sound—an image that indeed lies very closely to Freud's practice. Or maybe it is correct to say that psychoanalysis is simply sound in another form, provoking a personality that, just as the musician in Jankélévitch's scheme, nonetheless expresses "himself in the very act of not wanting to do so" (*Ibid.*, 42). Buried in this perpetual contradiction that music sets up are the dynamic outlines of the famous psychoanalytic patient—that *sound figure* who, via transference and the general mass of sonority of the spoken word, in the end expresses the absence of definitive meaning through a meticulous enactment of the communicative gaps themselves.

The production of sound within the confines of the psychoanalytic session presents us with a very interesting moment on which Freud himself relies inextricably: the externalization of the mute, unconscious element and its phonographic reassembly through the technique of acoustic amplification. And right away, what we encounter is resistance, the same one where "wishing not to express oneself is the great coquetry of the twentieth century" (*Ibid.*, 42).

Interestingly, Jankélévitch concludes that music, even though it cannot say anything apart from itself, does nonetheless possess the power to open up its own experience in such a way that we would be able to convert its latency into an estimation of a lived moment that had passed through it: "I have refused to grant music the power of discursive development, but not the power to trace experience of lived time" (*Ibid.*, 93). In this, music is very much ensnared in the thicket of latent, unacknowledged autobiography:

> This autobiography—if autobiography there is—is a bit dream-like: it conveys the "meaning of a meaning", that is, the meaning along with that which has revealed it, the secondary meaning, and it hides the primary meaning, the meaning that is in short absurd (because it is a mixture of sense and nonsense)—the very meaning of life cut off by death. (*Ibid.*, 57)[10]

Autobiography is a type of resonance that undermines its own certainty. It touches and sets upon the psychoanalytic mythology via the ear canal and the Greek legend of Echo, marked at each end by Narcissus and Oedipus.

ANNA IN THE KEY OF 'O'

Recent history of modern identity is more than familiar with Anna O.'s 'private theater', as she herself called it. The latter not only constitutes the beginning, the first presence of psychoanalysis before the formulation of its own concept, but also prestages everything that will eventually follow in the psychoanalytic practice. It is a dress rehearsal or a preamble to the psychoanalytic lexicon before the concept. Personal biographies and collective histories work in a very similar way.

Preambles (from Latin, *preambulus*, 'walking before') take strides before any actual displacement occurs; they are a conceptual treadmill of sorts. 'Before' in any case is more complicated than the simple preceding in time, because it can also be 'in front of', in the presence of an observer, both chronologically and topographically, when a space is actively shared, even if the body remains motionless. In fact, the intensity of such active sharing of space seems to increase precisely as the body experiences local paralysis, as psychoanalysis discovers through the use of the couch on which one escapes the field of vision. Freud, at one point, deliberately positions himself to the side, in this way facilitating an acoustic space of active listening. As Jean-Luc Nancy observes, a special type of presence is initiated here, not the standard "position of being-present", where we look at each other eye to eye. Rather, it is a presence as "a *coming* and a *passing*, an *extending*" through the ear.[11]

The interesting aspect of such 'before' is the question of ownership, which, formulated directly asks who possesses the experience and its biographical history? It is here that we enter the uncertain parameters of distances and the perpetually disputed definitions of art and science, psychoanalysis and music, and the individual personalities that are circulated through their systems. As we will learn momentarily, the ground on which such belonging rests, presented as the 'talking cure', 'examination', 'chimney-sweeping', 'recognizing work', etc., is shaky at best. All these formulations, later adopted by psychoanalysis more or less piecemeal, are actually invented—that is, *spoken*—by Anna O. during the spectacle of her 'private theater'—this condensed vignette of Vienna circa 1900.

Studien über Hysterie, written in 1895, five whole years before Freud's *Traumdeutung*, is a collaborative work with Josef Breuer. By the time of the second edition of the book, Breuer distanced himself from the general theoretical leanings of the work that by then were imposed on it by Freud. So the question of 'belonging' is present even here, because it seems that despite the two authors sharing equal weight and responsibility for the text (Breuer's name, following the semiotic convention that is to mark such equality, appears first), its message, in the end, belongs to Freud. This fact becomes even more intriguing if we juxtapose it against the story line of how Anna O. was 'discovered'.

The first important aspect of her discovery is that, from the start, she 'belonged' to Breuer. Anna was his patient, his case study—at least at the beginning, for, not being able to deal with her performances, with the creativity and inventiveness of her disorder, Breuer not only at one point invited an observer to witness them in hopes of gaining a second opinion but also, by delivering the results via a collaborative publication, inadvertently transferred the entire episode as much into Freud's archive, as it had ever belonged to his. At one point, then, through her inscription into the historical annals of psychoanalysis, Anna O.'s story changed the hands of 'ownership' and was attributed to Freud, thus mimicking on a different level the blueprints of a communicative modeling known as transference.[12] By now there is an entire history of such reenactments of formative encounters through the logic of transference between various registers of presence.

What originated as a space of accommodation between disease and its interpretation has been extended to the cultural fabric as such.

It would seem that it is precisely here where we can rest and stop delegating responsibilities, but we would again be disappointed, because the technique that initially became known as the cathartic method, as Freud acknowledged from the start and repeated over time in several places in his oeuvre (although with varying degrees of emphasis), was actually Breuer's. Freud simply adopted it, at first entirely, together with the application of hypnosis as a way of facilitating the 'talking cure', only later to give up this type of engagement for a modification based on suggestion (through verbal and physical clues, such as pressing on the patient's forehead), to finally settle on the method of free association—the backbone of *The Interpretation of Dreams* (1900), Freud's emblematic work.

The extent of this resonance of 'belonging' is not exhausted in that 'soap opera' whose three stars are Freud, Breuer, and Anna, for how can we lend our analytical ears to anything at all when sound itself defies clear provenance? Is its origin found in the source (even if that source is within us) or in our reception of its vibrating invisibility? The eye, for the most part, receives, cuts through, and penetrates; the ear absorbs into its vulnerability, and it is all gyration and dance inside. It is not an image (even though it often stimulates imagination) but an *echo* of reality. Hence, sound, and through it listening,

> is not exactly a phenomenon; that is to say, it does not stem from a logic of manifestation. It stems from a different logic, which would have to be called evocation, but in this precise sense: while manifestation brings presence to light, evocation summons (convokes, invokes) presence to itself.[13]

***

The case of Anna O. is fundamentally *musical*. It happens in real time, marked by the sonorous texture of listening, talking, and interpreting. All these are collected in a sort of phonographic repository that depends not only on recording but especially on the impulse for spontaneous production of sound as the expression of the crisis of form. And the most essential event in that entire network of 'belonging' is the act of composition, for ultimately everything that happens, from event sequencing to expression, is under Anna's direction. Breuer and later Freud have no choice but to be absorbed into the context and then replay it after the fact. Such reproduction and recovery are the only way anything theoretical can take place.

The question of utmost urgency occurs at the point of invention—the moment when information seeps through as the unintended consequence of overstretched identities. Freud's term for this process is 'failed connection' (*Fehlleistung*), translated into English as the very counterintuitive 'parapraxis' but also recognizable through the famous 'Freudian slip'. Although the issue here is not so much a slip in signification, given that the latter still occurs within the parameters of meaning, but a breakdown in the general mechanics of language, perception, and even kinesthetic action—where, for a brief moment, the frameworks of certainty are thrown off. This lateral motion permeates the psychoanalytic space and its setting to the core. It also presents a very apt, scaled-down model of Schoenberg's compositional technique, which is initially based on a series of *failed connections* between tones, especially those that the standard musical harmony has always been at pains to circumnavigate. In fact, Schoenberg builds an entire acoustic universe out of connections and relationships that were meant to be discarded and avoided.

Once again, all this brings about the question of the placement of identities, which operates under the principle of aural intensification and imposes on us the problem of sound and sonority. This question of possessing an identity circa 1900, of how, in the end, it might possess us, is a question that pitches, not only because it strays and hesitates but also because it rises above and beyond what its own configuration has the capacity to recognize, and that is why it becomes more than simply a question. In the process of repeated interrogation about genuine communication, this question itself becomes an *echo condition*, which from the start and at every subsequent moment perforates identities at their weakest points. As such, it not only matches the circumstances of sound in the uncertainty of its location but, most importantly from our perspective, also emerges as the *pièce de résistance* of modernity that asks about what belongs to it, and about its own sense of 'belonging'. All of this can already be surmised from the seemingly innocent, simply informative character profile of the first psychoanalytic patient revealed to us in the first few pages of Breuer and Freud's study. Let us now turn directly to them.

Anna O. is described as remarkably intelligent, well educated, of sharp intuition and critical understanding, with rich poetic imagination. All this makes her not very receptive to suggestion—a remarkable fact that, for now, we should simply keep in mind. She is a vibrant, sensitive young woman, prone to daydreaming, who displays an active imagination, toned down by quick and efficient reasoning. In Breuer's own words,

> She had great poetic and imaginative gifts, which were under the control of a sharp and critical common sense. Owing to this latter quality she was *completely unsuggestible*; she was only influenced by arguments, never by mere assertions. (SE 2: 21)

Of course, such assessment constructs a vivid contradiction in terms. It seems that Anna O. responds to the inherent impossibility of formulation into a name and an identity that would cover adequately the space of her unease (dis-ease). Hysteria, in its simplicity and all-encompassing reach simply will not do here, even though in the end Freud and Breuer will settle for the familiarity of its empty name, only now with newly redrawn boundaries. As Freud is at pains to point out, hysteria itself subsumes a whole range of other disorders. This fact makes it a much more complicated condition than the sanitized, medical presentation ever allowed for, rendering it practically useless as a diagnostic tool:

> I went on to consider the cases of neurosis which are commonly included under the diagnosis of hysteria. I reflected that it was not right to stamp a neurosis as a whole as hysterical because a few hysterical signs were prominent in its complex of symptoms. I could well understand this practice, since after all hysteria is the oldest, best-known and most striking of the neuroses under consideration; but it was an abuse, for it put down to the account of hysteria so many traits of perversion and degeneracy . . . Breuer's patient, Anna O., seems to contradict my opinion and to be an example of a pure hysterical disorder. This case, however, which has been so fruitful for our knowledge of hysteria, was not considered at all by its observer from the point of view of a sexual neurosis, and is now quite useless for this purpose. (SE 2: 258–259)

Whether Anna's disorder was to be placed under the historically tainted category of hysteria, or should have been classified under a different name, is a matter of mere nomenclature and of little significance from the perspective of our immediate focus.[14] On the other hand, there is no doubt that the 1895 manuscript, as well as Freud's later efforts, is an attempt to unmask the complicity of things and symptoms, remaking mere labels into conditions that respond to and are defined by a very specific technique of listening for the 'true' nature of a disturbance. This theoretical ear must be trained to contend with the muted presence of the patient, described as her unsuggestability.

But what exactly does it mean to be unsuggestable (a direct translation of the German *unsuggestibel*, which preserves much of its clumsiness in both languages)? The correct way to read this would be something like obstinate, not easily swayed or convinced, not prone to follow advice, inflexible. A more sensitive reading could reveal a graver affliction: a diminished capacity to focus due to uncertain dimensions, which are caught in the quicksand of constantly shifting shapes. The end result is the inability to define or even trace the frontiers of presence and absence. Anna O.'s unsuggestible characteristics decline medical or even theoretical intervention, thus revealing aspects that eventually become central pillars of the psychoanalytic theory: resistance (*Widerstand*), transference (*Übertragung*), and the presence of a 'parallel consciousness' that, like the bass line in music, underlies the entire compositional character of subjectivity.

Faced with Anna, Breuer is at the mercy of the performer/patient. He must accept what is given to him, what is presented as a mass of sounds, words, and actions, over whose sequence and content he has no control, even though at this early stage both Breuer and Freud still cling to the illusion of command by hanging onto nothing less than suggestibility of the entire setting, especially through hypnosis. Realizing the futility of this procedure, Freud will ultimately regain total direction over the performance by designing a stable slate of psychic history through the Oedipus concept and the general notion of sexuality. It is only in this manner that whatever escapes the patient can be handled by always binding it to this psychic drama—Freud's own 'private theater'. This realization also sheds light on Freud's steady opposition to any attempt at limiting or weakening the theory of sexuality at the base of his fully mature procedure. Without this point of attachment, this tonic and key that open the space already premarked, preformulated, and ready-made, the entire psychoanalytic setup would very quickly deteriorate into a cacophony of voices and their obdurate 'he said, she said' pseudo-communication. The Oedipus complex creates a very specific and nonnegotiable interval of intimacy. No matter what the patient says, he or she expresses only a closely situated relationship of agreement, related to a nonnegotiable event. Schoenberg will employ a very similar solution to the problem of dissonant sound on which his entire acoustic philosophy rests, where dissonant relations are treated as simply a more distant form of agreement, musically speaking. Recalibrating distances along these lines entails a fundamental change in perceptive dimensions.

***

Anna's hysteria was marked by general disorientation exhibited through the loss of language, especially its syntactic and idiomatic tonality. In fact, Anna invents linguistic resonance in her own key of 'O', which at one point enables her to conduct conversation only in English, use ungrammatical German, lose the power to understand her native tongue, and read only in French or Italian:

> There appeared a deep-going functional disorganization of her speech. It first became noticeable that she was at a loss to find words, and this difficulty gradually increased. Later she lost her command of grammar and syntax; she also no longer conjugated verbs, and eventually she used only infinitives, for the most part incorrectly from weak past participles; and she omitted both definite and indefinite article. In the process of time she became almost completely deprived of words. She put them together laboriously out of four or five languages and became almost unintelligible . . . For two weeks she became completely dumb and in spite of making great and continuous efforts to speak she was unable to say a syllable . . . Her paraphasia receded; but thenceforward she spoke only in English—apparently, however, without knowing that she was doing so . . . Nevertheless, she herself could still understand the people about her who talked German. Only in moments of extreme anxiety did her power of speech desert her entirely, or else she would

> use a mixture of all sorts of languages. At times when she was at her very best and most free, she talked French and Italian . . . She now spoke English and could not understand what was said to her in German. Those around her were obliged to talk to her in English . . . She was, however, able to read French and Italian. If she had to read one of these aloud, what she produced, with extraordinary fluency, was an admirable extempore English translation. (SE 2: 25–26)

Would it be a forced attenuation of facts to say that Anna used language musically? That she was not after meaning (a meaning that she resisted in any case, and which Breuer was forcing her to find) but form? That she combined various idioms into a simultaneous presentation akin to the immediacy of sound—a performance along the lines of extempore playing described by Eduard Hanslick, where origination and execution coincide and free the sonority of communication by allowing music and speech to enter into a liaison of spontaneous permeability? This would mean that the patient is speaking in a sort of speech-song, or *Sprechstimme*, Schoenberg's favorite technique of vocal performance. The half-sung, half-spoken method of *Sprechstimme* originated at the end of the nineteenth century. It was practiced by, among others, Engelbert Humperdinck in his melodramas. Schoenberg, in the course of giving instructions to singers, emphasized that "The melody in the *Sprechstimme* by means of notes is not, except for isolated exceptions that are specifically marked, intended for singing."[15] Rather, the performer is to reshape such 'speaking-melody' into a form that only suggests the indicated pitch, while making sure that it is clearly distinct from mere spoken word. Such singing that does not fall into a song and speaking without using speech are the type of hybrid assemblage symptomatic of the general profile of space and culture around that time.

This presentation of form, and curiously enough Anna's form, will find later expression in the formula of the famous psychoanalytic treatment, where the episodes of altering states of consciousness, loss of speech, translation, inexpressibility of meaning, absence of awareness, reconfiguration of reality through the performance of symptoms, catharsis, etc. will become the basic components of the psychoanalytic construction site—singular elements that repeat themselves in various sequences and inversions (like Schoenberg's twelve-tones) lifting the entire composition of a disorder off its moorings.

Breuer's encounter with Anna rests on the broad shoulders of expectation.[16] This is perhaps the signature of all treatment and every encounter with the 'face' of an illness, but expectation seems to take on new and monumental proportions around 1900, especially in psychoanalysis. It is the expectation for a return of the repressed, of being able to receive the history of a disease as payoff for even a minimal investment into a moment of engagement. This constructed moment requires all the experience, intelligence, and cunning of the analyst, who transposes disjointed impressions and sound bites into new figures of expression. Psychoanalysis does this by desynchronizing the immediacy of encounters and the 'meaning' of events. Thus, Anna O.'s performance is more than simply her 'private theater'—it is rather something akin to the genre of monodrama, where a single act stands in for all the complications and implications of the larger historical and psychic dramaturgy.

Anna O.'s pathological biography would eventually comprise, among others: absences (general dissociative states), various forms of limb and facial paralysis, deafness, loss of speech and language, periods of lucidity (normality), anorexia, bed confinement, misrecognition of the people around her, general unresponsiveness to the environment, and recognition of and responsiveness only to her therapist (Breuer). All these events belong to her, especially the moments of absence (as Breuer refers to them), or more plainly the states of self-induced hypnosis. This base condition of absence is caused by the secondary, intervening consciousness (what later becomes the psychoanalytic unconscious) that hemorrhages into and eventually takes over Anna's waking reality. Given this, the analyst has to do no more than to simply be present at the right time for the performance to happen. To facilitate this convergence between the patient and the analyst, their mutual presence is tempered by the correct calibration of personal frequencies. Hence, as Breuer points out, "I was the only person whom she always recognized when I came in; so long as I was talking to her she was always in contact with things and lively" (SE 2: 26). This contact and affirmation of presence took on an interesting tactile aspect: "She would never begin to talk until she had satisfied herself of my identity by carefully feeling my hands" (SE 2: 30). In this context of verification, where the doctor becomes merely an instrument in Anna's hands, any control that Breuer might want to exercise became not simply dubious but outright impossible. It is not he who sets the rules of the game but Anna, by conducting him into the scene like an overture to an opera. The analyst, in a similar fashion to a historian, becomes simply a channeling device. Like a magnetic phonograph needle he records the stream of sounds coming into the setting from the alternate consciousness. The ultimate aim is for the patient to find a way of utilizing this amplified archive, taking its instrumentality into his or her own hands, reassembling acoustic information into new autobiographical arrangements.

2 Psychology of Expectation

The most famous instance of the modern impulse to remake the world is embodied in Goethe's *Faust*. His act is singular and lonely: "He writes without consulting books, outside any discursive network."[1] What was still a half-fictional projection of modern culture's fate soon after became its material and psychological fulfillment, leading Pushkin, for instance, to observe that *Faust* is the "Iliad of modern life."[2] By the time the twentieth century arises from the soot of the Industrial Revolution, the expectation of almost *ex nihilo* artistic and technological creation becomes ubiquitous and matter-of-fact. It comes together with new subjectivity; a new type of identity snared in environmental contexts and their pathology, underpinned by the condition of expectations for *a psychology*—the kind of anticipation inscribed in weather prognostications, or the prognosis for an illness.

Expectations respond to the uncertainty of possession. They resonate with anticipation, but often are only the fulfillment of disappointment. Always operating under the directive of possibility, expectations can just as quickly undermine it, when the resolution of their potential turns out otherwise from what had been anticipated. Freud falls under the imperative of this 'cast-iron formula' of expectation, its flights of fancy and dynamics of feeling, quite blindly and unequivocally:

> I saw that my general prohibition has been ineffective and that I should have to take her frightening impressions away from her one by one. I took an opportunity of asking her, too, why she had gastric pains and what they came from . . . Her answer, which she gave rather grudgingly, was that she did not know. I requested her to remember by tomorrow. She then said in a definitely grumbling tone that I was not to keep asking her where this and that came from, but to let her tell me what she had to say. (SE 2: 62–63)

This exchange took place between Freud and one of his first patients, Frau Emmy von N. It became something very typical. Although at such an early stage (Freud analyzed her sometime in 1888 or 1889) all of this still appears quite naïve and even amateurish, the expectation of a correct memory and through it a *corrected* psychology will eventually find a much more sophisticated, subtle, and intricate expression at the heart of the psychoanalytic technique. But before that can happen, psychoanalysis must experience an early stage in its development, where the expectation to remember, and remember truly, takes the manifestation of a plastic and tactile form, in the most direct manner:

> My therapy consists in wiping away these pictures [of trauma], so that she is no longer able to see them before her. To give support to my suggestion I stroked her several times over the eyes. (SE 2: 53)

This technique of expectation or insistence, as Freud also calls it, is supplemented not only by an affinity towards the patient[3] but also by Freud's Faust-like infallibility:

> In these circumstances [of defence or resistance] I inform the patient that, a moment later, I shall apply pressure to his forehead, and I assure him that, all the time the pressure lasts, he will see before him a recollection in the form of a picture or will have it in his thoughts in the form of an idea occurring to him; and I pledge him to communicate this picture of idea to me, whatever it may be . . . Only in this manner can we find what we are in search of, but in this manner we shall find it infallibly. (SE 2: 270)

All this now sounds quite comical to us, pointing to a strange alchemy of conjuration, of extracting gold out of the impurity of traumatic metallurgy. And, in fact, even Freud discounts these early attempts at treatment. At the conclusion of his report on Frau Emmy von N.'s case, he appends an extensive footnote (one of many in the main body of the text) that begins with these words:

> I am aware that no analyst can read this case history to-day without a smile of pity. But it should be borne in mind this was the first case in which I employed the cathartic procedure to a large extent. (SE 2: 105)

But what happens later does not shed the identity of expectation and its particular type of resonance—it simply perfects and sharpens the procedure for its expression, displacing it onto the fertile ground of repression (*Verdrängung*). If expectation often disappoints in its outcome, repression is already a setback—a missed mark, a hindrance whose resolution is accomplished through an unmitigated and uncompromising demand to re-create its progress, retrace its trajectory. Hence, the problem is not a specific psychology whose validity rests on the particular and painstaking enumeration of contents, even when it seems like things operate that way. Rather the issue here is that of form, a psychological frame that delimits a base condition that must be owned up to, found, and acknowledged: the everyman's *nameless* psychology. Shoenberg's first monodrama, *Erwartung* (Expectation), is arguably the most brilliant expression of this psychological namelessness and its call.

***

Erwartung enlists the popular literary form of monodrama, already utilized by the Swedish playwright August Strindberg.[4] The monodramatic form instigates a fundamental shrinking of the world—a reduction into the minimal dimensions of just one act, whose confines are meant to capture an event in its intimacy by condensing it into a moment of explosive intensity. The result is a flexible space to which the subject is attached like a button to the mantel of a disproportionately exaggerated inner psychology. Schoenberg's musical treatment of this space enhances its basic qualities by externalizing what is considered the essential, expressive element of modern personality—emotionality, confusion, violence, and compulsion underpinning the psychological frame of reference. Not surprisingly then, the aim of Schoenberg's first stage work was "to represent in *slow motion* everything that occurs during a single second of maximum spiritual excitement, stretching it out to half an hour."[5]

Perhaps the best place to start a 'description' of *Erwartung*, if such a thing is at all possible, is at the end. The point of its termination does not present a resolution but maintains the suspense and tension of unfulfilled expectations. The work concludes indefinitely, with a series of sixteenth and thirty-second runs, trailing upwards accompanied by a shimmer of quadruple pianissimo. The woman protagonist, after a series of traumatic encounters that take her through a moonlit forest in search of her lover, ends where she began, with an endless search: "*Ich suchte* . . ." ("I was seeking . . ."). Apart from everything that can be said about *Erwartung*, the most crucial is the observation that it is a work of *analysis* in sound, one that, moreover, not only is inconclusive but also must go on, compulsively and obsessively. In this, it mimics Freud's approach to psychoanalysis that in its basic identity is also interminable (*unendlich*), even though practically, like Schoenberg's composition, it must stop. The potential locked in this inexhaustible repetition of the search, starting again in new configurations of the same elements, liberates mutation as a basis for psychology.

The monodrama, written rapidly over the course of seventeen days in 1909 and based on a libretto by Marie Pappenheim who might have been related to the famous Anna O. (Bertha Pappenheim), is a work that resists interpretation.[6] Its space is constructed on the basis of dizzying deflections of shapes and meanings. These run through a multitude of plateaus, such as the psychology of the musical form, the acoustic profile of the entire work and its story line, extending to such practical matters as performances of the opera, which had to wait until 1924 for its premiere in

Prague. In other words, the monodrama is mired in a nexus of displacements whose pattern, initially, even Schoenberg had trouble recognizing, as Egon Wellesz relates in 1921: "When Schoenberg showed me the score to his monodrama he said he could not explain how these short phrases were connected. 'I had to trust my hand' he said, 'like a painter who sets patches of colour which create the vision of a picture.'"[7] This is how Schoenberg's early encounters with his own sound take on the tactile form, the hand not exclusively the ear guiding the expression on paper, like Freud's pressing on the patient's forehead.

The analogy to painting is not surprising, as this is the period when Schoenberg took up this art, with some success.[8] Neither is the uncertainty of a nebulous identity inscribed into a work of art that searches out new phenomena: "Schoenberg says over and over again that a work of art must show a distinctive 'physiognomy'", relates Wellesz.[9] It is only in the 1930s in a series of theoretical drafts on the concept of the musical idea that Schoenberg will be able to delineate a physiognomy of sound that *Erwartung* embodied already in 1909.

Schoenberg's acoustic universe is like an organism, and quite literally so: "Above all, a piece of music is (perhaps always) an articulated organism (*ein gegliederter Organismus*) whose organs (*dessen Organe*), members (*Glieder*), carry out specific functions in regard to both their own external effect and their mutual relations."[10] Schoenberg's compositions are underpinned by the *plastic force* of repetition that will eventually find its most concentrated expression in the twelve-tone method, where sound is deposited into the intensified, multidimensional universe of sonic arrangements. The initial acoustic identity based on a row of notes is continually bent out of shape, inverted, split open, and sculpted. "Is it not true that repetition has an intensifying effect?"[11]—he asks. This is not just a rhetorical question. Schoenberg answers it affirmatively in print and sound. It has been observed many times that making and meaning are intimately connected in his artistic practice, to the point that anything that cannot be related becomes incomprehensible.[12] This relational nature of his musical thought process (and Schoenberg's sound is primarily that) is very similar to free association—the base through which psychoanalysis releases its power of interpretation as psychological *shaping* of seemingly free-floating, meaningless information. The most prolific and, in a sense, unconstrained period of free acoustic association is Schoenberg's early atonal phase epitomized in *Erwartung*. Later, the method of generating freedom through the acoustic analysis of sound, again in a similar gesture to psychoanalysis and its mature technique, will become more regulated and reliant on the 'authenticity' of a special method. This is why Theodor Adorno, who otherwise considers Schoenberg as the most progressive of modern composers, will ultimately mistrust the twelve-tone method of composition.[13] Schoenberg, on the other hand, in a similar manner to Freud will always maintain that the method is secondary and the body of sound, like the body of the patient in its disquiet, is primary: "We know that in the motive there must be present a certain unrest (*Unruhe*) that will give rise to further motion."[14] Thus, at least initially, in its raw form, sound is found in disquietude: "The explanation of the musical idea as unrest also illuminates why the development of music has taken the route of always inventing new kinds of 'unrest': the old ones could no longer agitate the nerves; stronger stimuli were required, and there developed an interest in discovering and experiencing how stronger states of unrest could be subdued, that is, how rest could again be attained."[15]

Erwartung is a study of unrest—this much is obvious, especially through its libretto. A woman, probably in a state of delusion, encounters a phantasmagoria of unfulfilled expectations. Searching for her lover, she comes upon a body in a clearing—it could be him, but is it? She finally rests on a garden bench, exhausted by her journey through the metamorphic forest. Perhaps all this was only a dream of blind wandering—this is not entirely clear, as the opera blurs the lines of division between reality and hallucination. It achieves this feeling after Schoenberg's minor but very strategic alterations of the libretto. Pappenheim's original was much closer to a clinical study of a disorder, probably inspired by the case of Anna O. and Pappenheim's own medical training that she had already put to poetic use, publishing four poems on the pages of Karl Kraus's *Die Fackel*, among them "The Autopsy Room" (*Seziersaal*). The ambiance and the story line of this stage work align perfectly with Freud's statement about hysteria, dating similarly from 1909: "When one carries out the psycho-analysis of a hysterical woman patient whose complaint is manifested in attacks, one soon becomes convinced that these attacks are nothing else but phantasies translated into the motor sphere, projected on to motility and portrayed in pantomime."[16]

Indeed, *Erwartung* is such a pantomime, on several levels of involvement, not the least on the musical one. It is not simply a representation of a disorder but its resounding echo. Disorders are lonely, isolating events. Each note in Schoenberg's score seems to carry the entire weight of this seclusion. Adorno observed that "Expressionism remains, in spite of itself, what art openly professed in the years around 1900: loneliness as style."[17] Schoenberg's monodrama, even in its genre designation, is an exercise in isolation. The notes, with no obvious connection, experience a type of sonic leprosy, growing and decaying in their solitary confinement. Adorno finds an especially powerful moment of this presence of solitude and desperation in the acoustic surroundings of measures 411–412, which accompany the words "thousands of people march past . . . I can't recognize you" (*Tausend Menschen ziehn vorüber . . . ich erkenne dich nicht*). In itself, this moment is an echo of Schoenberg's earlier composition, a song (op. 6) based on the poem by John Henry Mackay, which begins "Thousands of people march past/The one for whom I long, He is not among them!" Another, equally powerful instance of solitude and utter disappointment, where words and sounds come into condensed proximity, is a glissando in double bass (measures 70–71) from F down the scale to G-flat. It is a shattering moment of tumbling heads over heels down the ladder of momentary hope, right after the realization that the woman's lover is again nowhere to be found "*Aber du bist nicht gekommen*" (But you have not come). The work is full of such piercing and unfulfilled events. They are supported by unidentifiable musical structures, because the chords cannot be related harmonically in a standard manner. Squeezed into four short scenes, propelled forward by a protagonist who is herself without a name, they express the ultimately inexpugnable condition of anonymity.

But the question remains: how do all these notes connect? Schoenberg's expressed difficulty in understanding his own sonority has been replicated in almost all the subsequent analysis of the work. Certain patterns are identified, like the C-sharp-D-F figure that seems to be repeated in certain places throughout the work,[18]or the D-flat-C-A convergence appearing in various inversions in bars 248–249, "a dense network of relationships", as Carl Dahlhaus observes, that "cannot be dismissed as a chance accumulation."[19] But there is another way of approaching this perplexing, acoustic-material phenomenon of Schoenberg's sound that perhaps not surprisingly comes from the composer himself, who, with the benefit of time and reflection wrote in 1936,

> To symbolize the construction of a musical form, perhaps one ought to think of a living body (*Gesamtkörper*) that is whole and centrally controlled, that puts forth a certain number of limbs (*Anzahl von Gliedern*) by means of which it is capable of exercising its vital functions . . . If one were to presume, by way of comparison, that one such whole body could be the tonality of a piece of music, then the single basic tone, the fundamental one, would be relatively lifeless if it did not itself contain in its overtones those centrifugal and centripetal forces that make up its life and assign its organs (*Organen*) their functions. It lies in its nature to allow the forces that are unified and contained in it to develop and strive away from each other . . . Thus they become members (*Glieder*), thus they perform functions, thus they independently go their own ways, in which neither another member nor the whole body participates. Yet, thus, they must be also independent, able to detach themselves completely from the context and to prove themselves able to breathe and to lead an independent existence. If this comparison of music to a living being should thus far be applicable, then from the above discussion it must follow that the members emerging from the centrally controlled body (*Zentralkörper*) (whether it be either the fundamental tone or the fundamental tonality or a grundgestalt, motive, or theme) must be fundamentally different from it.[20]

In one sweeping gesture, almost breathlessly, we have here the characteristic *organicism* that animated the *fin-de-siècle* modernity. This kind of organic logic also connects all these notes in Schoenberg's compositions. It is a logic of developing variation, fundamental to Schoenberg's acoustic universe—variation deposited in the material of the body, musical and physical, that itself becomes groundless, given how its form has been loosened to freely associate things that should not go together. This aesthetics of dissipation, arguably best expressed in the medium of sound, is also present in other art forms such as painting and literature. Not unlike Robert Musil's *Der Mann ohne Eigenschaften*, we are dealing here with the relentlessly vanishing totality, a world without a clear resolution , without precisely delineated parameters, ingested by its own plastic force, and there is something primal and mystical about it all. This is how each acoustic shape—because this is what Schoenberg is creating, not mere notes—gains enough independence to act as its own center of gravity, becoming an *organ without a body*, thus rendering the body of tradition, in this instance tonality, inoperative. Schoenberg will always have a curious relationship with musical heritage—working from within it, undermining it from the inside, and releasing its transformative power as its own, previously unfulfilled potential. It is the virtuosity of this acoustic-plastic reasoning that glues all his notes together—the driving force behind the frame of reference in whose confines all the material sculpting of meaning takes place.

THE GROUNDLESS BODY

In the experimental 'echo chamber' circa 1900 the body's physical safety margins are dissolved. The expectations now reach beyond any reasonable limits, because it is the body itself that becomes the raw material in the dizzying search for new expression—one only needs to have a quick look at Schiele or Kokoshka.

Having given up the late Romantic rationality of emotional elevation (*Steigerung*), the setting of the *fin-de-siècle* modernity is no longer under the directive to maintain itself according to the well-established and clearly stated rules of philosophical interpretation of life and its bourgeois prescriptions for reassurance. Life is now burdened with the responsibility for new, progressively universal definitions that attempt to deal with the restless disquiet of a new century and the increasingly strained collective psychology . An attempt is made to stabilize the whirlpool of events by providing a life vest made of definitions that could make society buoyant again in the sea of fresh significations. Fundamentally, the aim is to produce a set of new values. The entire question of modernism, especially Viennese modernism, rests on the reinvention and discovery of these new registers of perception.

Fin-de-siècle in its 1900 edition is a curious blend of ambiguous and nameless forces, expressed most famously as the psychoanalytic unconscious. One is ensnared by experience, always hedging one's bets on the question of *accessibility* to a layer of reality that needs to find its voice. An attempt is made to unify experience and vault over the quicksands of the increasingly hollow presence. The indicators of this search for a powerful but indistinct presence are semantically inscribed into the *fin de siècle*'s *Ur*-formulations, as in *Ur-schrei* (primordial/primal cry), *Ur-trieb* (primal drive), *Ur-grund* (primal/primordial ground or condition), *Ur-mensch* (primal/primordial man/humanity), or even the *Ur-laut* (primal, indistinct sound).[21] All these are employed in the service of a new and unprecedented world view. Its topography at one point establishes a network of tattered intensities—the pitch of the entire discourse of the 'turn' (turn of the century, of art, music, psychology, etc.). Created is something like a map of modern sensory perception that measures experience according to what, at one point, André Breton described as 'slope-facts' and 'cliff-facts'.[22]

It is a world always askew, populated by events whose former trademark, the rationality of pure vision through philosophy and religion, has collapsed. And this in turn means that any search for a new form of expression has to move against the mechanics of standard, well-tempered, and calibrated systems of perception—musical harmony as tonality and the neat geometries of painterly perspective are the most obvious and powerful targets of irreverence. Instead, phenomena step over their safety margins, turning the comfort of old reassurance in common sense (*sensus communis*) into the anxiety of *individual responsibility* for 'meaning'. It is a rebellion against the mute coherence of everyday acquiescence, whose rage, in its first instance, is turned towards and against coordinated systems, including the deceptive and mystical technology of the organic sanctum, the 'plagiarism' of the factual truth it perpetuates.

This act that eliminates the body as the naturally endowed space of ethics and standardized truth is the most effective way of freeing aesthetic expression. One erases the center whose radius has always extended outward, only touching the circumference of the world at a point of more or less distant impurity. This is accomplished by turning the body into an aesthetic machine, where aesthetics itself becomes a special type of analysis that splits and fragments surfaces it encounters. This is the only spectrum through which the call for the 'new spirituality' or the 'new man' at the turn of the century can be understood. There are countless literary and polemical examples in these short few decades of Viennese and European modernism of precisely such a cry, initiated a few decades earlier by its philosophical patriarch—Friedrich Nietzsche.

***

It is no coincidence, then, that in one of his earliest works, Nietzsche deals with nothing less than the concept of history. As a recent invention that reconfigures the past through which modernity is relentlessly molded, history is full of possibility and danger. Its confines, try as one might, are not easy to navigate. Hence the title of his work takes on the outline of the entire argument: *On the Uses and Disadvantages of History for Life*. Nietzsche's message is one of caution and strategy. Given the propensity of modern culture to continually erect new idols, the past as history must be handled with care. Such a task has to be accomplished through the force that is from the start inherent to historical orientation—the force of plasticity:

> To determine this degree [of sleeplessness], and therewith the boundary at which the past has to be forgotten if it is not to become the gravedigger of the present, one would have to know exactly how great the *plastic power* of a man, a people, a culture is: I mean by plastic power the capacity to develop out of oneself in one's own way, to transform and incorporate into oneself what is past and foreign, to heal wounds, to replace what has been lost, to recreate broken moulds. There are people who possess so little of this power that they can perish from a single experience, from a single painful event, often and especially from a single subtle piece of injustice, like a man bleeding to death from a scratch.[23]

Thus, already in 1873, in one beautifully written paragraph, Nietzsche gives us the blueprints for the cultural environment at the turn of the century, including what later becomes known as the psychoanalytic theory. This exercise in memory and forgetting, their mutual reinforcement through plasticity is, in fact, almost point by point, the procedure used by Freud to handle his patients and their biographies. For treatment to be effective, the mass of shapes, anxieties, and events must be deposited in Freud's hands and sculpted. As late as 1937 Freud makes a specific reference to the binding power of the psyche, admitting that in certain cases "we are surprised by an attitude in our patients which can only be put down to a depletion of the plasticity, the capacity for change and further development, which we should ordinarily expect."[24]

Plasticity as an instrument that can be controlled and harnessed is isolated only by stepping aside, away from the organic space while preserving its elastic logic . It is a way of extracting the core of the body, its formula, from the physical and cultural limitations of the biological organism and the bourgeois processing of its confines—a way of discarding and making them insignificant as an archive of relationships beyond any challenge. That challenge now seemed possible, making the search for a new type of meaning move away from even the most sophisticated definition of 'man', because 'man', even the wisest of his species, is "a mere conflict and cross between plant and ghost."[25]

The radical forcefulness of the *Jahrhundertwende* modernity, its irresistible field of attraction as well as its extreme identity, finds placement in the search and expectation for a typology that would support the new form of the 'human' in its entire sensory spectrum—a version of Nietzsche's *Übermensch*. This is why it is also the age of the manifesto and the *expectation* lodged in its literary form.[26] But it is important not to reflexively fall back onto the standard assessment of that time, as simply the hour of great contempt and misguided calls for renewal. Rather, the project was the cultural and subjective renovation of immense proportions—the most profound modulation of the intellect and its psychology since the time of Descartes. For what is left after the human form passes through the condition of the 'beautiful-virtuous body', the 'evil-sinful body', the 'insignificant-administered' body, the 'split body', the 'docile body', the 'disciplined/technologized' body, the 'scientific body', and finally the 'purified and sexually sanitized' body[27]—after Wilhelm Reich cries out, "You don't have bodies, you are bodies!"?[28] What is left after this progression that takes place over the preceding centuries is perhaps only this: the subject *without* a body—an organic figurine, whose corporeality has shifted its mass towards the phenomenon of its own plastic makeup and its aesthetics that is extended to everything around. Could this be the reason why, for a short moment, art is put in the position of living in the everyday as its only currency, its use-value and meaning—the spirit of practically all the artistic/intellectual engagements at the time, from the Viennese Secession, onwards?

This reality of *plastic modernism* and its strained materiality is supported by sound, given sound's 'natural' elasticity. This is why its logic and the amplitude of its tone infiltrated almost every other mode of artistic and scientific expression, including psychoanalysis. It is not a surprise then, that Kandinsky, for instance, fell back onto the receptivity of sound to explicate the theory of color, where ultimately sonority and painting collapse into each other, finding themselves reverberating through the echo that deflects their identities, spinning them around, because one is now *expected to hear* color as well as see it—to hear visually, and to see musically. The point on the plane is silence, but "depending on the size and form, the basic sound of the point is variable."[29] Paul Klee attempted something similar, although he was much more cautious about the presence and the role of sound in his work, using it more as general inspiration than a technical device.[30] And of course, one should not forget Van Gogh, his influence on the Viennese artistic circles,[31] and his strokes on canvas that physically arrange paint in wave after wave of rippled, vibrating color. It is color breaking its own frames of reference. Never has the seeming silence of viewing been more acoustically driven—one only has to take a quick glimpse at Van Gogh's *Starry Night over the Rhone*, painted two years before his death, in 1888.

The process that decenters the subject also decomposes the body and its sensory machine. Such loosening is the fulfillment and the final acceleration of the erosion of visual perspective that gives up the worship of symmetrical form, including the one embodied in the notion of humanity:

> All philosophers have the common failing of starting out from man as he is now and thinking they can reach their goal through an analysis of him. They involuntarily think of 'man' as an *aeterna veritas*, as something that remains constant in the midst of all flux, as a sure

> measure of things. Everything the philosopher has declared about man is, however, at bottom no more than a testimony as to the man of a *very limited* period of time. Lack of historical sense is the family failing of all philosophers; many, without being aware of it, even take the most recent manifestations of man, such as has arisen under the impress of certain religions, even certain political events, as the fixed form from which one has to start out. They will not learn that man has become, that the faculty of cognition has become; while some of them would have it that the whole world is spun out of this faculty of cognition.[32]

To overcome the limiting horizons of presumed psychological completeness, an entire range of culturally endowed sensuality must be dismantled. The way towards the overcoming of this limit leads through the expressive modification of taste and perception. Aesthetics steps in and works with the purpose of rearranging the dullness and heritage of the senses. In this way, the entire project of modernism becomes a workshop for the production of perception that would be responsive and not merely reactionary—simply paying homage to intellectual and artistic education of the time gone by:

> He, who has taste, immediately says yes or no, before he himself knows why. He, who has taste, will be overwhelmed by pleasure or disgust, without being able to help it . . . [but] what we today call taste is merely comprised of [cultivated] memories.[33]

Culture as cultivation nurtures the root of the split between mind and body, thus always directing life towards the ghostly part, rearing 'men' who are informed by pure thought, apparently the most human of all elementary traits. As the prestige of thought advances beyond its humble beginnings, the body increasingly becomes simply an appendage, the phantom limb of reason.

But we should not be deceived here by well-popularized philosophical clichés. Even the famous Cartesian corporeality is not an absolute, opposite term to thought. In fact, the body is thought's symptom expressing the unwieldy nature of the organic mass that, in the end, does not lend itself to pure representation as a separate entity. Just as Freud will say that the ego is first and foremost the bodily ego, corporeality at that early stage of modern consciousness seems to be more like the *echo* of thought, a figure created by its own deflection that dissipates and returns to the point of origin, after the shape of the perceived "thing" had been registered. So "instead of the alleged exclusion of the ear from the search for truth, Descartes' philosophy enacts an uneasy truce between *cogito* and *audio*, a precarious entente between *entendre*, hearing, and *entendre*, understanding."[34]

This is why, reason or *cogito*, the way Descartes attempts to formulate it, is not the purely reductive, instrumental formula of scientific, representational thinking but a curious complex of presence through doubt that is affective as well as creative and thus the underlying 'cause' of the things immediately perceivable. The problem is not the antagonism between insufficiency of perception and the deeper, hence truer thought. The correct view is found from the beginning in intersections or their avoidance; in our ability not so much to conceive of an arrangement that is more perfect than what we are by necessity resigned to experience (although ultimately Descartes does fall back on the divine power filling in the gaps) but to experience a materiality that is fundamentally an expression of this *potential* that gives rise to everything else as it is later represented—a formula whose condensed form is found in this monument to modernity, *cogito ergo sum* (I think therefore I am). This famous 'recipe for being' of modern philosophy is not, technically speaking, a syllogism, because it is not a result of mediation where 'therefore' functions as the third term essential for mediation to take place. Rather, it is more like a temporal moment of fleeting thought overlapping with the mute material reality. That mute materiality (*res extensa*), nonetheless, is all vibration and variation of shapes, sizes, and qualities—their incessant extension. Descartes works hard to separate it off from reason or thought (*res cogitans*), because knowledge cannot be variable but must be unshakable (*inconcussum*).[35] He is not very successful in such an endeavor, coming under immediate criticism for doing so, because thinking does not give or deploy the 'I' but accentuates it as a trajectory already in place, amplifying its various components along the way. In this sense, reason (such as there is) is an element that is not preexistent or pregiven. Rather, it is something that must be actively maintained in perception (drawn as a line of thought, says Claudia Brodsky Lacour), or maintained like a long fermata note. Hence, Cartesian philosophy is not so much reflective or analytical as it is *productive*. Therein is found its radical modernity.[36]

Even though every extension of thinking is in itself imperfect, the fact that thought is applied at all and that it is able to think this imperfection, question it, and be suspicious of it in itself constitutes the only definition of perception. In Descartes's own words,

> Although it is not immediately apparent that so general a doubt can be useful, it is in fact very much so, since it delivers us from all sorts of prejudices and makes available to us an easy method of accustoming our minds to become independent of the senses . . . the mind, which in its intrinsic freedom supposes that everything which is open to the least doubt is nonexistent, recognizes that it is nevertheless absolutely impossible that it does not itself exist.[37]

This is a philosophy that *produces* perception through thinking, something, moreover, whose duration is always slightly out of sequence with the rest of the world:

> Just as signs for Descartes are not transparencies, his *cogito* is not an act of pure self-reflexivity. It is the production of "something" that, as a thing, is precisely not identical with thought: thinking and being are never conflated by Descartes into a single moment . . . In the *Discours* and in the *Géométrie*, thinking thinks into being what is not itself—a thing or a line—and that thinking already recognizes its own historicity: it is limited, not by an epistemic age to come, but by its own acknowledged occurrence "pendant le temps."[38]

Hence, the famous Cartesian dualism we like to reference almost instinctively is closer to being a complex sequencing of perception that provides the raw materials out of which experience and phenomena build their dimensions. In a very similar way to how Schoenberg will approach musical form, "the identity of the structure [in Descartes] does not pre-exist, but comes into existence with its presentation."[39] In this sense, the outside world becomes an extensive, Faustian *construction site*, not the archive of static, inherent knowledge. It is this making of the world through matter and its duration, through its fragmentation and divisibility, this spoiled part of objectivity, never aligned perfectly, that the *fin-de-siècle* modernity attempted to rediscover and make into another range of empiricism—one that is neither governed purely by the senses, nor falling back into abstract thought. This is why Freud considered dreams not as phantasmagoric illusions, but precisely another layer of experience and factuality, and why he eventually spoke of psychoanalysis as an art of *construction*, not interpretation—the latter being the label most commonly attributed to it.

What is then expected of the call for a new sense of perception that would finally unite the inner and the outer, without privileging the one or the other, is the only possible solution to the crisis of spirituality—a crisis Viennese modernism identified clearly and felt acutely. That solution is presented to us as the project of sensory refurbishment—an activity continually redefining the work of the subject where the physical and abstract body of tradition as an idealized necessity disappears. This was the only way to oppose the arid and futile divisions between thought and matter. And it was the acoustic quality of sound, in all its forms, that was able to accommodate the dissipation of these artificial divisions most effectively.

3 Play of the Senses

Modernity circa 1900 was self-effacing. A new sensory politics became prevalent, one that was not simply concerned with the administration but the *invention* of truth. This was not a politics inscribed into the great systems of bureaucracy and ponderous royal decrees that would be completely immolated in the muddy trenches of World War I, but a politics designed into the networks of psychological complexes. And in this sense, such a politics fell under a specific brand of aesthetics that tested reality by submitting it to relentless experimentation of how one could make it disappear.

To say that around 1900 society lost its sense of direction is to identify a symptom of a much more acute condition. More fundamentally, an attempt was made to redefine the sense and the idea of distance itself, including the human destinies distributed along its networks. The message, then, in all its sensory chaos was serious. One scoured reality for new relationships, along the way letting the worn-out ways decay. The intensity of this frantic search at its height compressed individual identity into an entire scale of psychological disorders. Freud invented, harnessed and then applied the stored energy of these psychological complexes, remaking them into a new form of analysis. At one point, they became something akin to a chemical agent—a 'rust remover' applied to the human psychological composition. Psychoanalysis relies on the premise of such a manipulation almost exclusively. It tries to dissolve resistance, arriving at a point where no defense is deemed necessary. Hermann Bahr summarized this character trait of the entire epoch to a point, when he referred to the modern self whose "main characteristic used to be resistance . . . today it is defencelessness."[1]

Defenselessness was reformulated into its own set of aesthetic procedures known as expressionism.[2] Expressionism arose in Germany at the turn of the century, maturing around 1910 and withering by 1925. It was a movement most strongly felt in painting and literature. Even though literary expressionism was practically absent from Viennese modernism, early forms of painterly expressionism (through Schiele and Kokoschka) as well as musical expressionism (through Schoenberg) are quite markedly present, with lines of influence running from van Gogh and Rodin in the case of Schiele, and from Strindberg as well as Dehmel in the case of Schoenberg.

Bahr provided an eye-witness account of its formation and, even more interestingly, a defense:

> Among the Viennese writers at that time Hermann Bahr was like a prophet of the new spirit. In all fields of art he helped and propagated the position which was called Sezessionistic [sic]. He encouraged youth and attacked convention; he was an Austrian who understood the European mission of his country . . . Bahr knew Schoenberg at that time from his Viennese performances and scandals. He helped him artistically and as a person. He had set up a foundation for the support of fighting artists, to which many rich Viennese gave their services and money.[3]

Not unlike Karl Kraus, Bahr assumed the role of the attention-rousing intellectual, although his analytical and polemical persona lacked Kraus's biting and often self-serving critique of opponents. In fact, Bahr was denounced by Kraus as the *Verwandlungskünstler* (the transformation-artist), for his seemingly unstable and morphing perception, a reputation that to some extent has stuck until today. All this, of course, does not discredit him as a rich source of raw information about the *fin de siècle* and its style of thinking. In fact, precisely the opposite is the case, especially when we consider that Kraus also searched for cultural redemption through dissolution of tradition-burdened, monolithic spaces. In Kraus's case, this stance not only resulted in his consistent and vehement opposition to psychoanalysis but also manifested itself as the relentless criticism of stable, collective identities in any form, including his own Jewish identity. As Robert S. Wistrich relates, "the young Kraus repeatedly voiced concern about 'Jewish characteristics' (*Eigenschaften*) which he feared might retard the *Selbstauflösung* (self-dissolution) that he resolutely advocated."[4] After all, one was supposed to be without characteristics (*ohne Eigenschaften*), to remind ourselves once again of Musil's formulation.

By the time *Expressionismus* was written in 1916, Western civilization was experiencing this type of dissolution full force, in the mud and putrid technological rationality of the trenches. Bahr's text reflects on the question of meaning, of finding oneself in the midst of existential disorientation, in the moment that bars accessibility to the logic of even the most mundane, everyday events—as was increasingly the case in the early twentieth century. And, at least initially, he resists the expressionist message, given how "I grew up with Impressionism" as he writes, and "I was an Impressionist before I knew of it" (*Expressionismus*, 10). This initial resistance is nonetheless overcome, in a large part thanks to the act of reflective introspection, in itself an ultramodern attitude: "You must learn to find yourself . . . new men are there, they bring the new time what it needs" (*Ibid.*, 11). What they bring the new time is nothing less than a new *emotive ground* out of which everything else should find a way of growing.

The operational manifesto for the new times makes the past itself vibrate according to a different range of frequencies:

> I prefer . . . simply to show, in what strange situation the clueless friend of art sees himself today, how this newest art affects him, what in it elevates him, why he believes himself threatened and what threatens him in it, and what it wants, why it wants it and if it does not, perhaps want something, what now must be wanted, perhaps even something, what has long been wanted, so that in this ultra-new the *arch-old* of humanity can once again be recognized. (*Ibid.*, 12, my emphasis)

We should indeed recognize Freud here—the Freud of *Totem and Taboo* with its primordial history, the Freud of *Civilization and Its Discontents* and its thesis of past repressions, and the Freud of *The Beyond the Pleasure Principle* with the primacy of the ancient death drive, as well as the Freud of *Moses and Monotheism* with its strange, fictional history of religion. All these texts aim to recover the heritage of modern culture through the reinvention of its memory, finding prehistory in hypermodernity. The result is that the tension between the beginning and the end, in a similar fashion to Schoenberg's *Erwartung*, remains unresolved, because "ultimately, the problem of the [twentieth] century is to exist in the non-dialectical conjunction of the theme of the end and that of beginning. 'Ending' and 'beginning' are two terms that, within the century, remain unreconciled."[5]

These long, almost endless cascades of beginnings and ends are so many reconstructions of presence that eventually become the archives of a universal psychology. This is not something unique to Freud but is the general adage of modernism, the thesis under which it operates. It feels compelled to leave a new imprint on the surface of its environment, by not simply *looking* to the past. Rather, the past is redesigned and transplanted into the present. The before and the after are brought together into a moment of interaction in an unprecedented way. This is done mostly through new techniques of perception, like psychoanalysis, expressionism, or modern, atonal music. Their copresence marks an intersection at which the present tests its limits. It persistently assaults the blind comfort of the everyday, releasing 'the now' in all its impulsivity.

In other words, one becomes defenseless through a process of relentless interpenetration, which *works at* reality and its design. Interpretation, at its most intense, releases the power of estrangement that recovers the unfamiliar past, both as individual memory and collective history. This is the only way to synchronize one's subjective presence with the seemingly chaos-driven timelessness of forces beyond one's control. And this is why the inescapable question that arises in all its vibrancy is, "*Was ist überhaupt 'echt'?*" (*Ibid.*, 33)—what is, above all, genuine? This question applies not only to art. It is intimately connected to the typology of the 'man' fit for the times, who must redress the entire expanse of reality. As we go on to reformulate and continually specify its parameters, it will become apparent that the dimension of authenticity circulates around the vascular system of the *fin-de-siècle* culture like a latent, genetic blueprint. It is an architectural extension of individual psychology, whose continual vitality depends on redesigning the senses. This means both the loss of their general biosocial characteristics and the development of new ones through displacement and dislocation, best encapsulated by Schoenberg, who at one point concludes that his 'music' can be heard only by a listener equipped with new ears for its reception—a theme of sensual rearrangement that characterizes much of what is at stake in Viennese and European modernism.

The question of the genuine experience accompanied by loss and redistribution points towards a considerable blurring of language. Speaking about expressionist painting, Bahr concludes that "It is not merely about art . . . [Expressionists] paint a new philosophy, a new religion . . . the release of humanity" (*Ibid.*, 39). The scale of this project automatically involves a linguistic crisis, and most of the setting circa 1900 fell under its burden. Its prevalence is an attempt to express something nonconventional and without precedent (*beispiellos*). One thus stumbles into a type of linguistic cavern, which in its murky surroundings must still remain communicative. If "obscure speech (*dunkle Rede*) angers the listener, he does not like to hear the clear one, or overhears it; that means, when truth is made too easy, too comfortable for him, then in turn he does not pay attention to it" (*Ibid.*, 45–46). Attention is the key to modern experience, especially the way one dispenses it, refocusing the sense organs, their culturally molded and biologically endowed organization. Ultimately, what makes all the difference is the correctly calibrated attention. It equips the subject with the ability to survey the world, to accost it in its symptom and one's potential release from it. And if we are to take stock of Freud and Schoenberg, then the most efficient description would perhaps be that all the complexity inherent in the oeuvre of both can be reduced to an attempt at finding and redefining the way one can and should *pay attention*—seeing oneself in the hermeneutic mirror of not only how messages are to be interpreted but also how they are constructed through the act of interpretation.[6] It is then not at all surprising that Bahr's next question, after reflecting what it is that we can call genuine, asks simply, "What is seeing?" (*Was ist sehen?*).

This question about the nature and the phenomenon of seeing paves the way towards what will later emerge as an interesting merging of the senses—the hybrid *Augenmusik*, or the music of the eyes. The road leading towards this very curious formulation is surprisingly straightforward, given that "the entire history of painting is always a history of seeing" (*Ibid.*, 51). But it is not a simple history, because seeing is not necessarily equivalent to vision, and this already implies a removal from the moment of perception as looking, discovering a unique ability that only art possesses—to touch a layer of depth beneath appearance. Thus, the quality of such a look is not simply sensory/mechanical but rather intellectual, releasing the apparatus of the eye that thinks from the 'autism' of mere sense impressions: "That is why the entire history of painting is also the history of philosophy, especially the unwritten one" (*Ibid.*, 51). This type of seeing proposes to resolve the worn-out, inside/outside split between the mind and the body:

> Two forces produce an effect, one upon the other, one outer and one inner, both basically unknown to us. By itself neither is sufficient. Appearance is created through both of these together. It is different for everyone, according to his own weaker and stronger interest, the activity of his eyes, the weight of his experience, the strength of his thought, the reach of his knowledge. When one of these conditions changes, every appearance must change with it. Most of the time man is not at all conscious of these conditions. But it can also happen that he feels them strongly then it can also transpire that he wants to change them. As soon as he becomes aware that his seeing is always an effect of an outer and his inner force, it becomes an issue of what he trusts more, the outside world or his own self. In the end, all human affairs are determined that way. As soon as he is at the point where he learns to differentiate between his own self and the world, where he says I and you, where he divides the outer and the inner, he has only one choice, either he flees from the world into himself or flees out of his own self into the world, or finally he remains at the border between the two; these are the three positions of men towards appearance. (*Ibid.*, 54–55)

The third and last of these subjective positions is the solution that the turn of the century modernism chooses in order to reform perception, because the task now is not simply to use what is given and decide in favor of mind or body but to *add* to the sensory machine, thus extending the senses in new and unprecedented directions. Anything less replicates the inflexibility of the old dualism, given how this type of adversarial system always takes as its starting point the impossibility of any reconciliation between perception and thought into account first and foremost. The ethos of the *fin de siècle* is to break through the glass ceiling of the old world differently. It considers the impossibility of an identity to reconcile itself with what faces it, whether socially or psychologically, as a false problem. Instead, elaborate detours are designed, continually deflecting the moment of confrontation when the subject and what Jakob von Uexküll, starting in 1909 described as *Umwelt* (environment)[7] "that island of the senses that wraps every man like a garment"[8]—come directly into contact, skewing the orbit of expectations. *Umwelten* (environments) are designed spaces that resonate to their own, unique frequencies—every organism possessing its own, unique *Ich-Ton* (intonation of the 'I').[9]

Hermann Bahr's diagnosis of how the modern individual ails through his classical past anticipates historical understanding of subjectivity, such as Foucault's, by decades:

> There comes the classical man, who, as Goethe says, "thinks himself one with the world and that is why he perceives the outside world not as something foreign that enters into the inner world of man, but rather he finds the corresponding images of his sensations in it. Since then, the entire history of the West merely develops such classical man. In the meantime, there always arise memories of pre-history that worrisomely threaten the classical evolution; it remains stronger. (*Ibid.*, 60)

The classical personality gazes longingly towards nature, saturated with the comfort of settled, even if often mysterious, possibilities. The classical look is

> the look of trust towards nature. Man turns himself more and more away from his innateness towards the outside. He becomes more and more the eye. And the eye becomes more and more perceiving, and progressively less negotiable. The eye does not have its own will any more, it loses itself in the stimulus, until in the end it becomes completely passive, nothing more than a *pure echo* of nature. Goethe still asked: "What is looking without thinking?" Since then, we have lived it. We can now answer his question and say to him, what it is: Impressionism. Impressionism actually is the completion of the classical man. What the impressionist looks for in seeing, as far as this is still somehow possible, is to separate everything that man can bring into the stimulus of his own accord. The impressionist [look] is that attempt, to leave nothing of man but the surface network. One is obliged to remind the impressionists that they do not "carry through" a picture. It would be more correct to say: they do not carry out seeing . . . and the impressionist mistrusts man, just like the pre-historic man mistrusts nature . . . he goes back to the first, initialization of seeing—he wants to see the stimulus in us at its first entrance. (*Ibid.*, 60–62, my emphasis)

Impressionism has been charged and found guilty because "it is the gaze of a time, which trusts only the senses, but is mistaken as to all the other powers of man" (*Ibid.*, 63).

Of course, in a contrary motion, it can be argued that impressionism retrieves the truth of vision by registering the raw imprint of light as it is really seen by the internal apparatus of the eye, thus giving us a more valuable, more 'information-nurtured' surface—its potential found precisely in the neutrality of impulses, about which a decision still has to be made. In this sense, such images would enable consciousness and the power of decision we normally do not experience (because in the eye all this happens involuntarily and instantly), giving us the option of manipulating space that normally is confined to an unconscious impulse of synaptic relays. This is how the possibility of inhabiting spaces technically speaking unavailable for encounters is created. Yet the fault of impressionism, its artificiality and deception, lies not so much in its technique but in the fact that it considers what it pictures as authentic separation, thinking that the latter can be simply reduced to a network of sense impressions that make 'man' stand out, apart from the world: "The impressionist . . . tries to uncouple appearance from the observer. The final result is: both dissipate" (*Ibid.*, 64–65).

The task, then, is "to come in the middle between Impressionism and Expressionism, to complete seeing, that violates neither the man nor nature, but gives both of them their due" (*Ibid.*, 67–68). Such middle is impure, and all the better for it. It allows crossbreeding between the senses to occur repeatedly. It calls for a new, hybrid condition of sensitivity that appends the senses to a new space, an alternate surface and field of attachment to the measured perspectives of cultural and biological geometries. The hope is that the entire sensorium can be regenerated in such a displacement—a sort of conceptually inspired and aesthetically carried out cloning that undermines the biological and the physical integrity of the body, without losing the critical margin of coherence it provides. And it is based on 'the music of the eyes' (*Augenmusik*).

The *retuning* of rationality that permeates the project of redesigning identity elevates aesthetics to the position of the 'cultural algorithm' and the key to universal relations. It employs the plastic force of modern events to extract 'truth' somewhere from below the skin of reality:

> this spiritual [mental] seeing, of which some people, and in childhood almost all, are capable, is more than mere remembering or a mere reproduction of sensory seeing. It is its own *production*, since the spiritual seeing has a creative power, the power to create the world according to different laws than the laws of sensory seeing. (*Ibid.*, 84, my emphasis)

It bears repeating that what is at stake here is not a reconstitution or even dissolution of the rational subject but in fact its resettlement. This is achieved by a series of topographical *dislocations* of thought, culture and perception in the decades surrounding 1900.

Such inhabitation of locations where 'things are no longer' is a corollary to Freud's famous '*Wo es war, soll Ich werden*' (where it was, I shall come into being). It is thus the case with Viennese modernism in general, as it is with Freud and Schoenberg that the symptom of the body as the classical look, or the traditional corpus of work based on overarching principles, such as tonality, will have to be overcome. One plays a game of high stakes, betting on Ernst Mach's *unrettbares Ich*, the irremediable, nonsalvageable 'I'.

Mach writes about the concept of the 'I' already in 1885, in his *Antimetaphysische Vorbemerkungen* (Antimetaphysical Observations). He advances a radical argument of a fundamental split in the structure of identity, something not considered seriously again as a psychological base for subjectivity until the 1960s and Jacques Lacan's psychoanalytic philosophy. The unity of the 'I', its objective consistency, is only a practical matter of convenient description. If things are considered carefully, we will realize that the 'I' comprises various, particular, and disjoined elements (just like the physical body) that give it a fragmented nature from the start—a condition for which there is no remedy, because the ego is never independent of the environment in which it is immersed. "The 'I' is not an unchanging, certain and sharply delineated unity."[10] It consists of partial elements, just like the material world is constructed from tones, colors, and various elementary particles. That is why "the 'I' has a low consistency, just like the body"[11] and why it cannot be salvaged—"*Das Ich ist unrettbar*".

We have an outline here for a radical and methodical self-treatment based on continually undermined connections (as will be recalled Freud calls these 'failed connections') between an identity and its expectations for a specifically embodied presence. Like "memory's wavering echo"[12] the subject becomes its own *recoding* machine, structured by a psychology that continually empties it of its cultural and biological contents, preparing its hollow figure for new perceptions. We will hear these echoes throughout and across the entire cultural spectrum. They are especially stark in Egon Schiele's painting where the body is submitted to something like a 'flesh eating disease', aesthetically speaking. Its physical parameters are continually fatigued, to the point that its physical coherence becomes increasingly difficult to save as a value in itself. Like chalk lines, its materiality is one innocuous swipe away from being completely wiped out.

Of course identity is still deposited in skin and bones. The difference now is how much weight they carry. The physical mass underpinning cultural psychology is for the first time calculated in units of resistance—a formula previously applied only to dead matter. Otherwise, everything should be experienced in the increasingly intoxicating state of weightlessness—an antechamber of this 'uncivilizing force' to which the entire setting makes a reference, a foyer to psychoanalytic unconscious that needs to be deciphered through listening, a waiting room for atonality as well as expressionism, for which one needs new ears. These 'new ears' are genetically reengineered organs, appended to the body. They are more than just a technological solution, or a prosthesis. Rather, these are the eyes that can also listen.

Bahr's summation of the encounter with the meaning of his times, in whose structure he is inextricably caught up, arrives at an interesting play with body parts:

> As soon as the waves of our inner life hit up to the eyes, we see our inner life, like we hear it, when its waves hit the ear. What, then, does all [effect] of music touch upon? To the tone-artist, the tones do not come from outside. He does not hear the world, he hears himself, the resounding of his soul in him . . . [And] what the painters of the newest movement want is, so to say, eye-music [*Augenmusik*]. (*Ibid.*, 104)

This musical orientation extends its own logic through sensory bioengineering and the act of inserting the ear into the eye because "man wants to find himself once again" (*Ibid.*, 110). This maneuver strips human sensory machine naked. It dissolves the body's traditional limits and negotiates the senses away from the body. And for the men and women of the *fin de siècle* who are learning to listen to their own humanity, this is the only way to live without submitting to aesthetic colonization by the bourgeois, impressionist guilt-machine:

> Never has a time expressed itself more richly and strongly as the bourgeois rule in Impressionism. The bourgeois rule was unable to bring forth music or poetry, all music and poetry of its time is always either an [emulation] of the past or an anticipation of the future . . . Impressionism is man's defection from spirit. The impressionist is the abased man who thus degraded becomes a gramophone of the outside world . . . the man of the bourgeois world does not carry through his life, he stops in the middle of seeing, for the man of the bourgeois world stops in the middle of life, just there, where the interest of man in life begins. (*Ibid.*, 112)

And this is why:

> Instead of the eyes, the impressionists still have a pair of ears, but no mouth. For the man of the bourgeois time is nothing as the ear, he hears the world, but he does not breath it in, he has no mouth, he is unable himself to speak about the world, to pronounce the law of the spirit. But the expressionist tears the mouth of humanity open again. (*Ibid.*, 113)

The mouth is pried open via the eyes that listen—a strange descent, a grotesque heritage of forced, accosted, compelled, and accelerated evolution. A play of the organs and surfaces—the mouth that is open, via the eyes that listen, hence also a production of sound, a scream: "Now necessity cries out: man cries after his soul, and the whole age becomes a single cry of need [emergency, *Notschrei*]. Art, too, cries with it, into the depth of darkness; it cries for help, it cries after the spiritual" (*Ibid.*, 111).

***

The entire *Jahrhundertwende* is characterized by such a cry. Futurists, for instance, cry and shout to the point of exhaustion. Perhaps the most famous representation of this shouting necessity is Munch's, although his came somewhat earlier, in 1883, anticipating all the ones that were to follow. Something very similar takes place in psychoanalysis, because there is not only the element that resonates out from the depths of the psyche, but also the indispensability of the voice, its clamor, desperation, indiscretion—of speech that yields to the compulsion of being outside of signification, screaming its head off. Schoenberg also uses this vocal power—out of all proportion. His compositions employ the voice more than any other instrument—a voice in the *Sprechstimme* style that often goes off the scale, not only in its pitch and volume but also in its coherence. It is a voice that tears itself apart as it escapes the body, and then shatters its own coherence again as it comes into contact with its environment—a disembodied voice that announces the time as a moment in dissolution.

All this coalesces into a new surface of engagement with the world that dispenses with proportionality, homogeneity, and continuity and explodes as a new symbolic form—a form that makes the space conform to the dimensions of the body, and not the other way around, where space is there as such *a priori* and identity must be always integrated into it in a dimension-conscious way. Given that it assumes a defining function, that it is a nexus of intensities, dynamics, and inflections in service of a physiognomic message other than its own physicality, the body and its voice create a condition of prolific instability. This is the only way that art is able to "produce life out of itself" (*Ibid.*, 114).

PLAY OF SURFACES

Invented during the Renaissance, the painterly perspective was an early, visual instance of activated expectation and predictability as a form of perception. Known as the *construzione legittima* (legitimate/correct construction) the technique attempted to counteract the contingency and distortion of vision. The formula that made space measurable by controlling the size of objects and their appearance as one's vision trailed off into the distance was by no means a commonsense view of the environment, but restricted to a group of maverick artists and mathematicians who marked distance by what is known as the vanishing point:

> For the modern vanishing-point construction distorts all widths, depths and heights in constant proportion, and thus defines unequivocally the apparent size of any object, the size corresponding to its actual magnitude and its position with respect to the eye. That is precisely the enormous advantage of the modern method, precisely why it was so passionately pursued.[13]

The vanishing point and the concept of infinity confounded with it invert the center of a visual field that, paradoxically, now is always beyond the horizon, out of reach. Such point of focus bestows a special, mobile identity onto objects as they move away from the eye's grasp. In other words, it is a center that, in the end, is not a center, as Jacques Derrida observed:

> It has always been thought that the centre, which is by definition unique, constituted that very thing within a structure which while governing the structure, escapes structurality. This is why classical thought concerning structure could say that the centre is, paradoxically, *within* the structure and *outside it*. The centre is at the centre of the totality, and yet, since the centre does not belong to the totality (is not part of the totality), the totality *has its centre elsewhere*. The centre is not the centre. The concept of centred structure—although it represents coherence itself, the condition of the *episteme* as philosophy of science—is contradictorily coherent.[14]

Only such contradictory coherence, very similar to Schoenberg's acoustic field, is able to channel dispersion.

The vanishing point corresponds to the point of view at the apex of the visual pyramid. The two are on the same plane, literally at eye level. This is why perspective, in its drive to make distortion predictable, solicits a response in the form of a subject—a response without which it cannot make any headway, in the most literal terms. The eye of the viewer is always suspended, so to speak, by the many lines that bind it to the harness of vectors in the field of vision; the subject is necessarily and inescapably inundated by the geometric reality, making identity into something like the "subjective metric"[15] element. And yet it responds to and connects with what, in the final analysis, does not belong to such a metric system. It is snatched away by infinity, rippling the smooth surfaces of perceptive symmetry. The subject thus becomes, in the assessment of Merleau-Ponty,

> a tacit, silent *Being-at*... the *self* of perception as 'nobody', like Ulysses, and anonymous one drowning in the world but who has yet to make his way through it. Nobody as imperception, evidence of nonpossession.[16]

The nonidentity at work in the setting of the *fin de siècle*, this 'nobody', is even more radical, for here the physical body and by extension the body of knowledge lose the coherence that visual perspective historically needed to maintain its operative parameters. To counteract these scrambled, standard vectors of linear perception, and to regain the communicative voice, new connections were attempted—for instance, by inserting the ear into the eye, as Bahr proposed.

In the long run, the symmetry and the organization of space that the linear perspective worked so laboriously to impose and maintain could not withstand the mystery of the vanishing point—the mysticism of its infinity. It is this point that always threw the constructive functionality of perspective into a state of disrepair, finally making it collapse under the elegance of its own construct. Thus, the element that was meant to be only a part (albeit an extremely important one) of the whole, the element that was to be working towards maintaining a stabilized field of vision, became itself a point of value, desire, and contention. We can speak of the moment of *recoil* here, because what used to be the vanishing point became once again enclosed in and ingested by the body, creating a vanishing dimension of depth that eventually became known as the unconscious. The principle of its operation moves along similar lines, but the consequences of that movement are profoundly different, for what 'transpires' when the horizon of visibility is inscribed in the body is nothing less than the event of vanishing of the body as such, exemplified by the recalibration of its sensory input and output. By the same token, what happens in psychoanalysis is simply giving *voice* to this inverted vanishing point. In this manner, we find ourselves once again at the threshold of the music of the eyes (*Augenmusik*)—a strange contraption indeed, but an unavoidable one because the eye, in order to advance beyond its own limitations, must learn how not to see so that it is able to perceive anything at all.

The notion of subjectivity as a point of view arose during the Renaissance along with the technique of linear perspective. We can find clues to this mutuality of purpose in the word itself, which, from *subicere* to 'place under', a combination of *sub* 'under' and *jacere* 'to throw', evokes all the connotations of being subjected and subjugated to, of being subjects in the political sense, on display, under investigation or surveillance. In this way, one also becomes the subject matter, the universal library of histories and experiences, their orthodox knowledge and potential corruption that must be known, studied, and uncovered. Perspective is a slice of space one chooses and pays inordinate attention to. It engenders subjectivity because it not only imposes a certain notion of viewing but also presents what is only a spatial fragment as universal topography, seemingly revealing space in all its naked attributes. Both Freud and Schoenberg, each in his own way, rebel against its seductive comfort.

***

Alois Riegl was arguably the greatest art historian of the *fin-de-siècle* generation. Riegl's *Spätrömische Kunstindustrie* (1901), for instance, was a book to which Walter Benjamin ascribed the greatest influence on his thought.[17] The list of *fin-de-siècle* intellectuals who elevated this work to the status of a leading paradigm of history is indeed long—among others, it includes architects Walter Gropius and Peter Behrens, as well as the sociologist Georg Simmel. In 1922 one of Simmel's most famous students, Georg Lukács, named Riegel as perhaps the most important nineteenth-century historian.[18] In fact, Riegl invented from the ground up what in time became known as the history of art, remaking it into a fully defined field of study by reformulating the dominant paradigm of art's evolution proposed by Gottfried Semper. In Semper's view, the study of art should be primarily concentrated on the technical aspects of creative work, such as techniques of production and materials, excluding the broader network of cultural connections and significations. Riegl's argument expressed through the concept of the artistic volition or will (*Kunstwollen*) displaced this technical aspect of artistic production from its leading position, at the same time bringing the multifarious, creative network of art's constitution as a total phenomenon to the forefront.

The discourse of *Kunstwollen* is lodged firmly in the interval spanning material reality and its abstract definition. Riegl argued that specific modes of artistic production cannot provide the ground that defines the characteristic profile of an art period. On the contrary, artistic contexts must be approached through a method that stands apart from the intricacies of technical practice. On another level, this formula pointed to the crucial difference between the paralyzed, symmetrical bodies of knowledge and practice and the animated, anthropomorphic corporeality of aesthetic perception. A genuine response to the environment is possible only when art is able to acquire individuality of expression, and the self, by perceiving honestly, becomes an aesthetic creation.

Riegl proposed to his contemporaries that the history of visual arts be divided into three general periods based on divergent perspectives or modes of seeing, underpinned by the mechanics of distance management. In a series of lecture notes published posthumously as the *Historical Grammar of the Visual Arts*, the issue of historical perspective on art is presented along a three-pronged spectrum of: (1) *the near view*, encompassing the voluminous heaviness, symmetry, and tactility of Egypt; (2) *the normal view*, encompassing the subtle balance between naturalism and idealism found in ancient Greece; and (3) *the far/distant view*, which reflects the early Renaissance modernity—its focus on optics, proportionality, and evenly distributed vision in the point of view/vanishing point dichotomy. If painterly perspective implicated the body in interpretation of space by relating it to geometries of seeing, then early twentieth-century modernism turned this formula right around. It banished the body itself into the vanishing point instead of thinking of it as the motionless summit of visual orientation that only looks longingly towards the horizon. Modern art and modern identity are examples of how space has emancipated itself from its earlier transcendental confinement. This is how "the formlessness of modern art . . . stresses the infinity of space stretching into depth."[19] To understand this better, we might want to consider Erwin Panofsky's engagement with the problem of 'historical optometry.' Panofsky's 1927 study *Perspective as Symbolic Form*, apart from its purely academic message, in itself constitutes an early response to the *fin-de-siècle* modernism, as it indeed follows closely in the footsteps of Alois Riegl.

Painterly perspective represents a special case of cognition. This is the power of its symbolic form. By stabilizing the idiomatic and unwieldy organization of space inherent to the untrained eye (the preperspectival vision), geometric perspective recoils from the psycho-physiological dimension in favor of mathematical space. This kind of recoil, and with it, passing off the responsibility for visual representation from the self to a system of administered vision, is cast in terms of aesthetic versus theoretical space. Aesthetic space is multidimensional and frayed at the edges. Theoretical space collects phenomena into the stabilizing but also enervating coordinates of symmetry; it flattens the image of what we see, making it consistently available for contemplation and viewing. Describing ancient Greece, Panofsky observes that

> The art of classical antiquity was a purely corporeal art; it recognized as artistic reality only what was tangible as well as visible. Its objects were material and three-dimensional, with clearly defined functions and proportions, and thus were always to a certain extent anthropomorphized. These objects were not merged in painterly fashion into spatial unity, but rather were affixed to each other in a kind of tectonic or plastic cluster . . . the represented space remains aggregate space; it never becomes that which modernity demands and realizes, a systematic space.[20]

At one point, this dynamic-responsive topography becomes eroded. Around the time of the Renaissance, space becomes translated into systematic, mathematical models. The three-dimensional, vital space of the ancients appears as the two-dimensional surface of modernity; the individuality of bodies and objects *as* space is removed in favor of the predictability of their movements *in* space, a shift that in itself, as Panofsky observes, is not without its own benefits.

The creation of the modern subject can take place only through this act that abrogates the independence and responsibility of objects in space, their individuality. Symmetries of vision arrange things according to proportional relationships of distance. No matter what point of view is taken, one is always certain of seeing the predictable arrangement. Whereas the ancient Greeks, through their psycho-physiological 'perspective', *create* space as the expression and consequence of bodily presence that must be accommodated, the geometrical perspective partially immobilizes that presence by giving space a priori qualities of homogeneity, seamlessness, and smoothness, thus clothing the body irrevocably in its dimensions.

Panofsky's analysis, not without its own problems, differentiates between the ancient and the Renaissance perspectives, a line of distinction that already arises in Riegl. It suggests that the Greek visual field, at least in part, was more intimately calibrated. Based on the curved as opposed to the linear spectrum, the Greek eye apprehended the world at an angle (Euclid's eighth theorem) and not through the monocular-linear point of view. Even though this seems to be a tenuous theory, this belief in the authenticity of sight and its loss falls in line with Bahr's and his contemporaries' call for the reinvention of vision that would directly invigorate or augment cultural activity, and not merely instruct in seeing. "In any case, I hate everything that merely instructs me without augmenting or directly invigorating my activity," wrote Goethe.[21] Panofsky, writing in the 1920s, is much more cautious and indecisive about this last point:

> the perpectival view, whether it is evaluated and interpreted more in the sense of rationality and the objective, or more in the sense of contingency and the subjective, rests on the will to construct pictorial space, in principle, out of the elements of, and according to the plan of, empirical visual space (although still abstracted considerably from the psycho-physiological "givens"). Perspective mathematizes this visual space, and yet it is very much *visual* space that it mathematizes; it is an ordering, but an ordering of the visual phenomenon. Whether one reproaches perspective for evaporating "true being" into a mere manifestation of seen things, or rather for anchoring the free and, as it were, spiritual idea of form to a manifestation of mere seen things, is in the end little more than a question of emphasis.[22]

Regardless of whether one reproaches Panofsky for trivializing and underestimating the consequences of such a 'matter of emphasis', one thing is indisputable. The responsibility for vision now falls onto the subject, who must carry the linear perspective within his or her conceptual apparatus. In this sense, the world and its contingency, a contingency that the subject instinctively resists as he carries the visual cone in his eye, disappear. It is this sort of underestimation that the *fin-de-siècle* modernism rebels against. The slogan of this rebellion is the freeing of 'man' from the shell of symmetries. This maneuver rests on the idea of art that, like Schoenberg's music, *recoils* from harmony that entraps sound in homogeneity and proportion. Primacy is given to *organic* surfaces liberated from the constricting uniformity of organization, whether it is the physical body, the body of knowledge, or tradition.

One does not have to look that long to find an intriguing correspondence between Western musical harmony—its symmetric sound based on one fundamental tone (the tonic)—and painterly perspective. This much was surmised already by Schorske, who observed that "tonality in music belonged to the same socio-cultural system as the science of perspective in art, with its centralized focus."[23]

Tonality prescribes the rules of *legitimate construction* defining how pure, organized, and pristine sound is to be produced. It standardizes distances between notes and preempts distortions, engineering a set of provisions not unlike the ones found in geometric perspective, setting specific, invariable increments of spatial progression for the benefit of coherence at all cost and at first sight. In this sense, classical harmony in music creates something akin to *visual sonority*, based on the idea of *perspectiva* ('seeing through'). The *fin de siècle*, especially Schoenberg's 'music', employs the opposite premise by deregulating proportions and distances, and releasing wave after wave of musical figures, what Schoenberg refers to as *gestalten*—shapes and floating arrangements of sound. This is because "above all a piece of music is (perhaps always) an articulated organism whose organs, members, carry out specific functions in regard to both their own external effect and their mutual relations."[24]

There is nothing more unpredictable and liberating than the disembodied 'organ'. Freed from the operational imperative of mechanical systems, it is now saturated with the anticipation of action (performance), expressed through the plastic, malleable materiality—a space that is produced by divesting normality of its coherence. In this sense, we could describe this creative tension in the manner of Freud's transference (*Übertragung*)—a space that cracks under the high voltage of information it attempts to carry. A new type of *responsive* materiality is thus created as something that can be continually fine-tuned and sculpted. This is how the body of the psychoanalytic patient, perhaps the artistic, psycho-physiological body par excellence, is slowly immersed in the pre-signifying, aesthetic communication known as the analytical session. The effect is a space of overlapping intensities, where the messenger and the message become one form—a form that is not rational/symmetrical but imbued with *distortion*. Distortions express the moment of possibility in its full intensity. Schoenberg's 'music' is populated almost exclusively with these moments of creative distortion, where cohesion without harmony predominates.

Perspective, the practical invention of its presence as a historical variable, which by and large takes place in the context of turn-of-the-century Vienna, initiates a line of questioning and a series of assessments that prove to be crucial in apprehending the positioning of the subject, in both senses of the word: the subject as the ego and the subject of modernity. What transpires here is a certain play of space and surface, or we can say in one breath with Riegl, using his grammar to describe the history of visual arts, the play of the *organic* and the *harmonic*:

> Man's artistic activity . . . continually fluctuates between two poles: the *harmonic*, which strives to bring the immutable formal law of crystallinity to view in all motifs, including organic ones; and the *organic*, whose supreme goal is to represent organic motifs in all their accidental and transitory momentary appearances . . . [and] we arrive at the following observations.
>
> *a*. Any worldview that values natural things only in physically improved form will generally be predisposed toward the harmonic, for physical beauty is equal to harmony, that is, symmetry and proportion. Conversely, a worldview that strives exclusively for the spiritual improvement of nature without regard for its physical *habitus*—or that simply tolerates the transitory appearances of nature as such—will be intrinsically inclined toward the organic.
>
> *b*. In periods when utilitarian purpose constitutes a work of art's sole reason for being, people always prefer to make motifs as

> harmonic as possible, or, as we shall call it here, to *harmonize* them (*harmonisieren*); conversely, during times when art is an end in itself, people tend to make them organic, or to *Organize* them (*organisieren*) . . . As for particular purposes, the harmonic generally corresponds better to the decorative and practical functions, while the conceptual purpose naturally tends to require organic solutions.[25]

The immensity of this project leads to the boldly stroked framing of history, presenting it almost ready-made for consumption and further manipulation, a similar type of engagement Freud practices—one only needs to remember here the key to cultural history and personal biography, the story of Oedipus. Modernism at the turn of the twentieth century is taken by this sort of view into the distance that, ironically and not unlike the vanishing point, attempts to bring things closer while assigning them to the somewhat nebulous realm of primeval memory. In the case of Riegl, this sort of manipulation takes on the shape of two cultural fields of action that at any given moment assume responsibility for the shared reality by 'organizing' (*organisieren*) or 'harmonizing' (*harmonisieren*) its variable dimensions. Spun out of this coupling of organization and harmonization are the stitches of the type of argument Freud and Schoenberg will be able to make through psychoanalysis and atonal acoustics, respectively.

These two concepts, of course, present almost immediate translation difficulties. The contrast between 'organization' and 'harmonization' drawn by Riegl loses its distinctness in English, given how harmonization could simply be interpreted as a specific type of organization, and vice versa. Yet the problem is that Riegl works with a line of demarcation between the *organic* and the *inorganic* and their perpetual expressive interchangeability, the cycles of their fluctuations throughout the history of visual arts, finally concluding that "our examination of the course of art history yields a picture of a fiercely progressive organism continually reined in and steered along by harmonism."[26] The somewhat cumbersome translation of the German noun *der Organismus* as the English 'organism' would probably be better rendered as 'organicism'.[27] In any case, it is a key term to understanding not only Riegl's thinking but also much of what comes after it, in the form of Viennese modernism, especially Schonberg's *organic sound*—a characteristic the composer becomes conscious of only gradually, with the analytical benefit of the passage of time.

What are the limits of aesthetics based on the 'organic quality' of permeability and fudged contours between the inside and the outside, the near and far, sound and silence? These are the operational nodules of perception and its exasperation so characteristic of the turn of the last century. Thus a plane is created on which 'man' becomes the echo of his organic constitution.

This kind of elementary constitution further amplifies Ernst Mach's assessment of the ego as an impermanent complex, put together by the ever variable combination of materiality and sense impressions. Thus, "that which we so much dread in death, the annihilation of our permanency, actually occurs in life in abundant measure."[28] Riegl's assessment of modern art and culture also makes this process of dissolution the centerpiece of the entire process. Coherent and clearly distinguished individuality, still visible in Dutch portrait paintings, "is merely an echo of the now-obsolete notion that a person is truly an individual entity."[29] On the other hand, modern individuality is paradoxically a collective, social expression of continually developing relations, like Schoenberg's sound eventually becomes the effect of perpetually developing variation. Thus in 1899 Riegl noted that

> Today art has other objectives. It must depict not the individual, who no longer exists—or, more precisely, is a set of molecules too tiny to be depicted—but the universal connectedness of all natural phenomena. This art presents its subject not as a physically unified individual but as a complex of optical appearances that strike the beholder's eye instantaneously. You might be thinking that this is just the exaggerated theory of a crazed hypermodern man. But you will see things differently if you trace the history of portrait painting back a few centuries . . . What is the external purpose of a modern portrait? Portrayal of individuality; the patron, at least, still desires this. What is art's purpose? Portrayal of causal relationships in nature. Extrinsic purpose and art's purpose are thus two separate things . . . art's purpose will completely subsume the external purpose and people will cease having their portraits made.[30]

This instantaneous vision that apprehends everything at once in a network of continually developing variation while making individuality disappear is nothing less than Bahr's 'music of the eyes.' It is therefore not a coincidence that Bahr devotes a laudatory section to Riegl during his discussion of expressionism—the power of Riegl's thought, according to Bahr, is perhaps outstripped only by Goethe in *Farbenlehre*.[31]

***

Oskar Kokoschka, one of the few unswerving supporters of Schoenberg,[32] in a gesture similar to that of Schiele, explored and liberated the material logic of the body's own physicality and the mystery of its opacity. Such mystery is enclosed in the elasticity of its physical constitution that gives us the stored psychology of not only the human psyche but also space in general. Taken at its lowest intensity, the logic enclosed within the shell of material phenomena is by default nothing else than the logic of decay. There is no better expression of this logic than sound, especially the sound in the shape of Schoenberg's music. This is also why the entire fin-de-siècle reality, from its shouts for the primitive, especially its rediscovery of the Ur-schrei (primordial call), through the flowing lines of Jugendstil, or even the bent-wood furniture of the Thonet brothers,[33] to the primary system of drives in psychoanalysis and the sounding of the unconscious, becomes in a large measure acoustic—the carrier of a misrecognized principle, theory, and method of resonance and the principle of its manifestation through distortion and dispersal. "The cry tends to exceed all language, even if it tends itself to recuperation as language effect."[34]

4 Acoustic Perspectives

In 1634, at the threshold of modernity we are intimately familiar with, Marin Mersenne, considered by many observers 'the father of acoustics', proposed a hypermodern formula of merged sensations. Music is "the perspective of the ear, like perspective is the music of the eye."[1] Two hundred and fifty odd years later, the *fin-de-siècle* sensibility embraced this type of resonating perception completely. Expression became a 'conversation' of frantic, de-ranged, and fleeting connections. Culture ceased being a matter of civility and cultivation, becoming an instantaneous echo of direct encounters with the physical structure of the world.

In this type of environment history ceases to be merely expired time. Much more soberly, it becomes a *method* for the recovery of memory, always in need of adjustment, so that it may reach the status of the leading, foundational event—a formula that Freud makes the centerpiece of his psychoanalytic design. In this manner, both history and individuality acquire a new form of corporeality through the collective past. Such past becomes a *voice*, increasingly more direct—a record of culture sequenced into the rhythm of appearance and disappearance, their form and surface: "Form and surface always sustain a certain relationship; and although it is true that surface can sometimes outweigh form, there is no such thing as a non-corporeal work of art, or pure surface."[2] Form and surface, two generic components of any identity but most intensely felt in aesthetic production, are volatile elements—engaged in a contest (*Wettschaffen*) that resolves itself into a periodic compromise. This compromise is a form of creative production we know as a world view (*Weltanschauung*):

> In art, man re-creates nature as he would like it to be and as it indeed exists in his mind . . . This impulse, which is in fact identical with the drive to create art, derives from man's striving for happiness. All human culture can ultimately be explained in terms of this striving . . . Consequently, man creates a vision of nature in his art that frees him from nature's instability; he imagines nature to be better than it looks . . . The comforting view of nature is something man creates in his mind. It affects man's relation to every object in the world without

> exception. Thus it entails not only the relations of man to extra-human nature, which we call an understanding of nature in the narrow sense, but also the relation of one person to another, which we call the understanding of morality. We can pull all this together under the single term "worldview."[3]

This creates a pattern of engagements leading to the already discussed principles of artistic production based on organization or harmonization. But, more importantly, the increasingly dense and overwrought contents of art will finally reach the self and inscribe themselves onto the surface of individual psychology like grooves into the phonograph record, thus creating a type of identity able to ask about *self-expression*. It is an identity that struggles to remember and is always caught in the network of deceptive chronologies—an individual who yearns for reconciliation between the cultural surfaces and the rawness of psychological depth, which always somehow drives him or her beyond the momentary legitimacy of various complexes.

Legitimate construction is an expression of the cultural necessity to measure and stabilize distances, not only in terms of history but especially in terms of the social, political, and physical dimensions. If classical psychology was always created after the fact, in support of a certain world view, at the turn of the twentieth century this process was reversed. Psychology became the elementary dimension—its horizon cluttered with events whose main function was to bring history and the individual into the tension of an altogether different encounter based on the contingency of relationships, not their predetermined trajectories.

The attempt to measure and calibrate the topography of the world through symmetries of predictability was of course not given up completely. This type of engagement reemerges again and again in the elaborate, organic paradoxes of *Jugendstil*, for instance, or the forcefully broken-up linearity of cubism, where attempts are made to stabilize the volatility of the incessantly receding surface of smooth, classical reason. Such mixed topography also found expression in the grotesqueness of symptoms etched into the surface of the body of the *fin-de-siècle*'s now iconic neurotic personality. Just as these were meant to be genuine expressions of the new 'harmonizing' world view, it was also the case that the direct reactions against such surface reality—for instance, in the guise of Adolf Loos's 'ornament and crime' adage, or Freudian psychoanalysis, that is, reactions of depth and its 'organics' against the surface—claimed for themselves the standard of a genuine expression of understanding and living. This is how perspective as a way of seeing, now deposited into the human body and physical materiality in general, acquired the general form of *displaced information*.

The tension between depth and surface splits the *fin de siècle* and later the entire twentieth century in two, while all along fuelling the desire to regain the completeness of form and presence:

> In the twentieth century, the shared law of the world is neither the One nor the Multiple: it is the Two. It is not the One, because there is no harmony, no hegemony of the simple, no unified power of God. It is not the Multiple, because it is not a question of obtaining a balance of powers or a harmony of faculties. It is the Two, and the world represented by the modality of the Two excludes the possibility of both unanimous submission and combinatory equilibrium. One simply must decide.[4]

This decision is continually *deferred* until around the middle of the century, when it takes on the shape of ideological politics of exclusivity. Until then, the desire to be one while remaining double, a condition that in psychoanalysis is expressed as the famous interchange between the conscious and the unconscious elements, is so overwhelming that it can be stitched together only by a new form of plasticity. More often than not, such plasticity is located in the voice of the subject who speaks from his or her own depth, arriving at a momentary equilibrium, where identity becomes a pronouncement, an *evocation*, not a manifestation or demonstration. It is this model of pronouncement and evocation that later in the century will be utilized and distorted by the politics of ideological formations.

In the recursive moves of continual try-outs, any identity deposited in this hesitating space of obstinate, double meanings must be continuously transplanted, from one to the other, and back again. Driven by dissention, decadence, and discomfort, but also the hope for a recovery, whether of memory or some type of authenticity, the senses are inevitably manipulated, accelerated, and organically modified to be harvested as new types of perceptual arrangements. All this, of course, is a consequence of the new, defining immediacy in which the *fin-de-siècle* identity finds itself. The latter is an encounter with the limit and its specific type of world view as total *reorientation*—its passion to be the one-word, one-dimensional answer to the prevalent existential vertigo. There is an entire series of such answers, which repeat themselves in various inflections: symbolism, primitivism, expressionism, Fauvism, Cubism, futurism, Bruitism, Dada, surrealism, psychoanalysis, atonality, and so on, well into the century. They find their limit precisely in their desire to be the code to a reality that flees towards the horizon without a coefficient of proportionality. Eventually, the experiment becomes the immediate exhaustion of limits right in the moment of perception, opening another plateau of experience—"something like a rotting on the spot [and] a nourishing decomposition."[5]

***

In the thirty-fifth introductory lecture to psychoanalysis, "The Question of Weltanschauung,"[6] Freud takes on the concept of world view directly. The definition is all-encompassing and expresses the consummate spirit of the times:

> *Weltanschauung* is an intellectual construction which solves all the problems of our existence uniformly on the basis of one overriding hypothesis, which, accordingly, leaves no question unanswered and in which everything that interests us finds its fixed place. It will easily be understood that the possession of a *Weltanschauung* of this kind is among the ideal wishes of human beings. (SE 22: 158)

After proposing to formulate the issue in such a broad manner, Freud immediately turns to the question of psychoanalysis and concludes that "the answer as regards psycho-analysis is made easy" (SE 22: 158). The easiness of this answer, it seems, comes almost too laconically and dismissively to Freud:

> As a specialist science, a branch of psychology—a depth-psychology or psychology of the unconscious—it is quite unfit to construct a *Weltanschauung* of its own: it must accept the scientific one. But the *Weltanschauung* of science already departs noticeably from our definition. (SE 22: 158)

This departure centers on the issue of desire or ideal wishes—an important component of the original formulation. Science, according to Freud, is not simply responding to reality but is an active agent in its construction. It operates in the margins of 'truth'. As such, it is realistic about its own power and acknowledges the variable degree of proximity built into its constructs, admitting to limitations, deceptions, and errors.

The central theme of Freud's text on *Weltanschauung* is the legitimacy of perceptive environments. One can distinguish among a number of possible systems of perception. Predictably, one of the most challenging, serious, and potentially damaging of such all-encompassing realities is religion. Freud's dismissal of art as a viable option is as interesting as it is surprising. According to this argument, art does not have the power to inflict indelible and grave enough consequences for it to be considered in the same light as religion or even philosophy—that is, as a system of impressions that has the ability to impose its own mode of perception on reality in general:

> Of the three powers [religion, philosophy, and art] which may dispute the basic position of science, religion alone is to be taken seriously as an enemy. Art is almost always harmless and beneficent; it does not seek to be anything but an illusion. Except for a few people who are spoken of as being 'possessed' by art, it makes no attempt at invading the realm of reality. (SE 22: 160)

That art is far from harmless and can indeed be utilized to invade and control perception should have been clear to Freud by the 1930s, when the lectures were delivered—only a cursory look at the relationship between, say, Italian fascism and the artistic avant-gardes such as futurism would have been enough to invalidate such a claim. It is indeed a long-standing and curious question as to why Freud always and quite forcefully insists on pulling psychoanalysis by its hair, so to speak, into the direction of a scientific paradigm, especially given his admission that psychoanalysis lacks one of its essential components—the cause-effect experimental setup, an omission that does not go unnoticed. This, for instance, leads Wittgenstein to conclude that

> The difference between a reason and a cause is brought out as follows: the investigation of a reason entails as an essential part one's agreement with it, whereas the investigation of a cause is carried out experimentally . . . It is a way of speaking to say the reason was subconscious. It may be expedient to speak in this way, but the subconscious is a hypothetical entity which gets its meaning from the verifications these propositions have. What Freud says about the subconscious sounds like science, but in fact, it is just a *means of representation* . . . [where] as in aesthetics, things are placed side by side so as to exhibit certain features.[7]

Psychoanalysis, then, is a type of *aesthetic arrangement* of information arrived at by juxtaposition of elements that are not obviously related. As such, Freud's depth-psychology lies somewhere in the transition zone between art and science. The issue is more than just one of legitimacy, because the latter would be better addressed by insisting (and this, in the end, happens anyway) on psychoanalysis's own uniqueness, on its special epistemological underpinnings based on constant *discursive hesitation*. This is what gives psychoanalysis precisely the element Freud wants to deny—the construction and perpetuation of its own world view. Why then play up the cover-up?

Dealing with psychoanalysis through the mitigating circumstances of science removes much of the responsibility for its own identity from its own hands. The motivation for such a displacement arises together with the characteristics of Freud's fully mature theory. At issue is the system-wide consistency, where theory and technique match their frequencies. Science presents the best cover because it instigates the process of morphing with its own method—an act very similar to how the ego expresses its psychological identity. Moreover, this elevation of scientific spirit is also an attempt, at least partially, to mute the aesthetic element in the psychoanalytic practice. Falling back onto the crutches of science counteracts the moment when the analytic session is overpowered by the white noise of the incessantly streaming consciousness of the patient, which, at any one point, can become parasitical, feeding on the instrument of its own analysis. Freud refers to this situation as transference, or more accurately a type of high-intensity transmission (*Übertragung*). This is also the greatest achievement of *Weltanschauung*. And, it is the same kind of frequency that underpins the grand and illusory world-building characteristic of individual pathology. In contrast, science, even though it is much more uncertain about its own stance, is always firmly placed within the worldly parameters:

> It is true that it [science] too assumes the *uniformity* of the explanation of the universe; but it does so only as a programme, the fulfillment of which is relegated to the future. Apart from this it is marked by negative characteristics, by its limitation to what is at the moment knowable and by its sharp rejection of certain elements that are alien to it. (SE 22: 159)

In other words, science already operates in the 'indentation' of a basic repression that situates it firmly in the space of the reality principle and its 'truth'. Any idealism of unfulfilled wishes is integrated chronometrically, through the distance of time, into the overall setup as something to be addressed in the future—that is, displaced onto the horizon of desire's vanishing point. This is how science becomes aware of its limits. Through this argument, Freud doubles his systematic elaboration of ego-psychology, which, at one point, becomes the synonym for psychoanalysis as such—a mouth that can now speak its own hesitation, and not simply respond in the language of certainty, like it does at first.

Essentially, there is nothing more scientific than the voice that tests reality instantaneously, while modulating the self into a form of a continually *delayed* presence. Unlike vision, a scream into the void returns immediately as its own echo. A gaze refers the subject back to itself as a foreign body. Only listening has the power to claim identity back as that self, even if in a distorted form.[8] And if the involvement of the eye is at one point necessary, its effectiveness is found through the apparatus of the *Augenmusik* and its sonar-like vision. Science and the modern individual find their voice, for the most part, through each other. They become more and more refined through simultaneous and ever expanding engagement.

THE EGO CHORUS

The blueprints established in the early days of Freud's practice find consistent elaboration throughout his career. Their echo is still clearly heard towards the end of Freud's creative life, in the 1930s, in such texts as the *Dissection of the Personality*, which builds on perhaps the most significant conceptual venture in Freud's theoretical output, *The Ego and the Id*. All the sources are consistent: the ego is in charge of maintenance and resolution of world views. As such, it mitigates, judges, and discerns impulses coming from three equally powerful directions—external reality, the superego, and the unconscious id:

> We are warned by a proverb against serving two masters at the same time. The poor ego has things even worse: it serves three masters and does what it can to bring their claims and demands into *harmony* with one another. These claims are always divergent and often seem incompatible. No wonder that the ego so often fails in its task. Its three tyrannical masters are the external world, the super-ego and the id. When we follow the ego's efforts to satisfy them simultaneously—or rather, to obey them simultaneously—we cannot feel any regret at having personified this ego and having set it up as a *separate organism.* (SE 22: 77, my emphasis)

The near impossibility of reconciling the three contradictory channels of communication, each operating on its own frequency, manifests itself in the form of the perpetual background noise of anxiety. As it becomes elevated to a high enough pitch, a pathological condition ensues. Nonetheless, the ego or the 'I' (*das Ich*) is continually engaged in its task of taking care of knowledge and perception, deciding and very often coming down on the side of error as to how such knowing should be accomplished and what conclusions drawn from it. As such, it is very much operating acoustically, through cognitive dissonance.

Dissonance is only a looser form of agreement or consonance, as we learn from Schoenberg, who works towards the elimination of this negative sound category.[9] Even after Helmholtz's scientific explanation of sound based on mutually resonating overtones, it became increasingly accepted in modernist musical circles that approaching dissonance as a clash or a form of disagreement that must be resolved has in the end nothing to do with any 'natural' characteristics of tones. In fact, until the seventeenth century and the reconceptualization of harmony by Jean-Phillippe Rameau, who applied a mechanical model to acoustics (something that earned him the sobriquet 'Newton of the arts'), sound was perceived as a fluid continuum.[10]

Such artificially instituted, sharp division between sonic agreement and disagreement was consistently put to the test and eroded throughout the nineteenth century, until its complete elimination in the so-called atonal music, around 1900. In this form, dissonance takes on the identity of only a more distant, related element, disposing of its false, negative identity. Harmony is not necessarily agreement, not even in the traditional acoustic spectrum. One only needs to think of the contrapuntal polyphony in Bach. Harmony as agreement is making a step similar to the visual solution of distortion—it is an imposition of a symmetric perspective onto sound that stabilizes unpredictability.

In a similar fashion, the psychoanalytic ego like the atonal sound that was being forged at the same time must find a way to embrace dissonance. This is the only way it can claim the unique role of being in the position of the coordinating element, actively tempering and retuning the many impulses coming from the cultural environment and the *acoustic unconscious* depths. This is how the always strained ability to adjust is extended beyond its own limitations. The modern individual is presented with the basic and undeniable truth of physical experience that for the first time threatens its fundamental, biological constitution. The ego must find a way of responding to this threat, while also remaining responsive to inner compulsions. The solution is to implement the method of composition, assemblage, and arrangement as new plateaus of topographical orientation. The ego, in fact, becomes a finely tuned amplification mechanism. Like a work of art, it is able to coordinate the contradictory realities of the three acoustic streams into the chorus of simultaneous expression, as it also responds to the "ear of conscience" (SE 19: 57). As its own *echo-complex*, the ego achieves this through the artistry of sound, mixing "an intensive spacing of a rebound that does not end in any return to self without immediately re-launching, as an echo, a call to that same self."[11]

Structurally, this line of reasoning is integrated into the system of Freud's depth-psychology. By making the ego creatively independent of the three masters it serves, one is able to mold it as a *separate* construct. The solution might seem cosmetic, but Freud and psychoanalysis need and absolutely depend on it. To counteract the pressure of being squeezed in between the impinging forces of the unconscious and the superego and then assaulted from the side by external reality, the ego is now given its own dimension of space with its own containment field. This happens by integrating the three streams of information into the ego's own physiognomy as so many points of contact that act in a manner similar to Schoenberg's never given-up definition of his method of composition as constantly developing variations of elements related only to each other. There is no imaginary focal point here, such as tonality in music or the vanishing point in painterly perspective, that could complete the design.

All this depends on inventing a new type of distance—something that must be activated together with the constant danger of complete degeneration:

> In its attempts to mediate between the id and reality, it [the ego] is often obliged to cloak the *Ucs.* [unconscious] commands of the id with its own *Pcs.* [preconscious] rationalizations, to conceal the id's conflicts with reality, to profess, with diplomatic disingenuousness, to be taking notice of reality even when the id has remained rigid and unyielding. On the other hand it is observed at every step it takes by the strict super-ego, which lays down definite standards for its conduct, without taking any account of its difficulties from the direction of the id and the external world, and which, if those standards are not obeyed, punishes it with tense feelings of inferiority and of guilt. Thus the ego, driven by id, confined by the super-ego, repulsed by reality, struggles to master its economic task of bringing about *harmony* among the forces and influences working in and upon it. (SE 22: 78, my emphasis)

As a separate organism, the ego becomes that "rotting spot" of "nourishing decomposition"—a hybrid form that actively translates, reworks, and reassembles all the impulses that shoot through it like through a vibrating speaker. Modern identity has always been a lot weightier than just a psychological impulse—its traditional image. This is why, according to Freud,

> the ego is first and foremost a bodily ego; it is not merely a surface entity, but is itself the projection of a surface. If we wish to find an anatomical analogy for it we can best identify it with the 'cortical homunculus' of the anatomists, which stands on its head in the cortex, sticks up its heels, faces backwards and, as we know, has its speech-area on the left-hand side. (SE 19: 26)

All these divergent paths cutting through the organ of the ego and the methodological procedures of healing the psychic forms that are always ready to be shattered, like crystal (SE 22: 59), impel Freud to propose an analogy that would in some measure capture the 'truth' of identity's operative principle:

> In thinking of this division of the personality into an ego [das Ich], a super-ego [das Über-Ich] and an id [das Es], you will not, of course, have pictured sharp frontiers like the artificial ones drawn in political geography. We cannot do justice to the characteristics of the mind by linear outlines like those in a drawing or in primitive painting, but rather by *areas of colour melting into one another as they are presented by modern artists.* (SE 22: 79, my emphasis)

This is indeed a curious but not completely unexpected expression, even taking into account Freud's initial statements about art as a cultural force that, as we will remember, dismissed it almost out of hand. Freud's writings are replete with such paradoxes and contradictions. And one cannot help but wonder whom Freud could have had in mind when he referred to the 'modern artists'—perhaps someone like Kandinsky, whose visual paradigm rests on a sophisticated theory of color based on the dynamics of sound. "Art is almost always harmless and beneficent; it does not seek to be anything but an illusion" (SE 22: 160). This statement stands in a stark contrast to the analogy drawn between the structure of the psychic apparatus and modern painting, especially its aesthetic effect of blurred origins and dimensions. It also contradicts the entire paradigm of psychoanalysis.

The problem of art is buried in the application of the psychoanalytic technique itself—its technical, theoretical, and cultural components. As might be suspected, any strict distinction among these three moments of psychoanalytic operational identity is already a distortion of its overall shape. Not unlike the modern painting Freud calls on for didactic help, these moments of self-expression intermingle in a thoroughly fluid manner.

About the only place where such a line of demarcation makes some sense and is defensible relates to the question of art itself, and it is perhaps the only way to grasp Freud's very convoluted statements about it.

The most important of such statements comes some twelve years before the paper on *Weltanschauung*, in a short essay entitled *Formulations on the Two Principles of Mental Functioning*, written in 1911. It is within the confines of this short article that Freud delineates his theory of pleasure/reality principle, and its structural integration into the progressively elaborate, complicated, and intriguing modeling of the human mind.

Building on his startling clinical observation of neurosis, Freud is "now confronted with the task of investigating the development of the relation of neurotics and of 'healthy' individuals in general to reality, and in this way of bringing psychological significance of the real external world into the structure of our theories" (SE 12: 218). We can anticipate by now that the entire setup rests on movement between realities, and the strategic play of perceptions. Most interesting here is a series of compensations and partial resolutions, of which art will be one, and the fully mature ego, the most sophisticated and efficient compensatory mechanism that operates in a very similar way. All this, in its progressive incarnations, falls under the structure of the reality principle:

> The state of psychical rest was originally disturbed by the peremptory demands of internal needs. When this happened, whatever was thought of (wished for) was simply presented in a hallucinatory manner, just as still happens to-day with our dream-thoughts every night. It was only the non-occurrence of the expected satisfaction, the disappointment experienced, that led to the abandonment of this attempt at satisfaction by means of hallucination. Instead of it, the psychical apparatus had to decide to form a conception of the real circumstances in the external world and to endeavour to make a real alteration in them. A new principle of mental functioning was thus introduced; what was presented in the mind was no longer what was agreeable but what was real, even if it happened to be disagreeable. Thus setting-up of the *reality principle* proved to be a momentous step. (SE 12: 219)

The distinguishing feature of the reality principle, then, is its ability to incorporate *dissonance* into perception.

In a very ambitious account, all the more so considering the brevity of the discussion, Freud proposes to recover the evolutionary chain of this dynamic exchange between the pleasure and the reality principles that eventually leads to the unique recalibration of the sense organs into a new type of psychology. Only such a maneuver makes room for consciousness that is enabled through the mechanism of attention (because now reality must be observed) along with its functional operative modes of notation/memory and passing of judgment, of deciding what is true/

false, what agrees with the world and what does not. That the judgment most frequently comes down against the impulses of the pleasure principle, because these are always in some way in the position of irremediable challenge to reality, is the primary effect of culture and society. It is also at this point that action as appropriate motor discharge takes over. Crucially, and in a counterintuitive manner, abstract thought according to Freud is but the highest stage of repressed motor activity: "Restraint upon motor discharge (upon action) . . . was provided by means of the process of *thinking* . . . It is essentially an experimental kind of acting" (SE 12: 221). But even at this point further adjustments are necessary, because the drives of the pleasure principle in their clamor for attention prove to be continually swelling. The final adjustment, then, is a split in the thinking apparatus itself. Only this lowers the intensity of the struggle through a tear in its own fabric:

> With the introduction of the reality principle one species of thought-activity was split off; it was kept free from reality-testing and remained subordinated to the pleasure principle alone. This activity is *phantasying* [sic], which begins already in children's play, and later, continued as *day-dreaming*, abandons dependence on real objects. (SE 12: 222)

This independence from visible materiality comes very close to and in fact *overlaps* with the identity of sound, its acoustic-material structure, and its resonant effect upon reality.

The chain reaction proposed by Freud is similar to a chord progression or, to be more accurate, progression of discord. Every step accomplishes a change of state, a condition of reorchestrated relations. This is the mechanism that governs Freud's world view. It is also related to the general progression of life, its emergence from the mute, chemical stage incorporated in *Beyond the Pleasure Principle* (1920) into the theory of the death drive. In fact, in his *Autobiographical Study* (1925) such transformative procedure is presented explicitly:

> the depth-psychology revealed by psycho-analysis was in fact the psychology of the normal mind. Our path has been like that of chemistry: the great qualitative differences between substances were traced back to quantitative variations in the proportions in which the same elements were combined. (SE 20: 56)

If we think of Ernst Mach and the material fragmentation of the self, or someone like August Strindberg, who continually worked with the theme of organic transfigurations, or even Émile Zola, who professed the procedure of dissection as the writing tool, we can thus locate Freud firmly within the logic of the *fin de siècle*—its existential, scientific, philosophical, and artistic vocabulary.

Even if it could be said, after all, that Freud's science is only a technique of analysis, analysis is the frame of reference through which the *fin-de-siècle* culture finds its coordinates. One leads an analytical life, a life in need of dissection, an existence that must continually invent increasingly elaborate procedures for the most accurate description of how things come to be, how they are accomplished and maintained. This is the shape that the space of the fully mature ego ultimately takes in Freud's setup—that is, it is primarily an ego that probes and tests reality, the 'I' that stands at the crossroads and hears voices. Placed at the intersection of the three information streams of external reality, the psyche, and culture, the ego ingests the vanishing point. The ingested vanishing point instead becomes a *point of audition*. Such a point is the structural core of depth-psychology that unleashed a form of analysis as the *aesthetics of disappearance*.[12] This is the case not only because the ego works over and recomposes whatever comes its way but also because it is in constant danger of losing coherence, of not being able to deal effectively with proportions and distances, with their distribution and maintenance:

> Actually the substitution of the reality principle for the pleasure principle implies no deposing of the pleasure principle, but only a safeguarding of it. A momentary pleasure, uncertain in its results, is given up, but only in order to gain along the new path an assured pleasure at a later time. (SE 12: 223)

The ego is in charge of this safeguarding mechanics through nothing less than the surface of a world view and its actualization.

Not surprisingly, then, Freud also elevates science to a more privileged position, albeit a qualified one: "It is *science* which comes nearest to succeeding in that conquest [of the pleasure principle; its safeguarding]; science too, however, offers intellectual pleasure during its work and promises practical gain in the end" (SE 12: 224). We have already encountered this during the course of our discussion. The interesting question is the one topographically in the middle: what lies in between the two principles, what kind of space is being created via this method of safeguarding and analysis, of paying attention and through it, reality testing; what kind of perspective does it engender? We have come full circle here, and returned to the problem of art and sound.

THE ART OF HEARING AWRY

The contest between art and science seems to be without a resolution, given how each is the desired limit of the other. Schoenberg, for instance, has no compunctions about science wanting to be art, and art's need to be scientific:

> The difference between art and science lies herein: that even where both aim to represent the same area, science must try to include all conceivable cases, whereas art confines itself to those that are characteristic, appropriate, or otherwise "fitting." While science will therefore have to place every case in the clearest light, art may change the relationship of meaning to heighten their effect . . . If science provides facts that it orders according to common principles, art produces images in which facts are freely joined to common principles, so that the sense of what is to be stated can be clearly grasped and at once . . . It is clear that a difference in the manner of presentation and in the principles of construction must also reflect this basic difference and that in art the meaning of what is called logic has to be somewhat modified, even though fundamentally the human mind is capable of only a single manner of thinking.[13]

One universal thought becomes two styles of thinking—the artistic and the scientific. Both are in a position of weakness, because they miss a part of themselves in having become what they are. Nevertheless art has the liberty of willful creation that is not given to science. This is why psychoanalysis and Schoenberg's mature method of composing with twelve tones occupy the middle space, wedged between art and science—the two never stop desiring to be one—the fundamental logic that permeates the *fin-de-siècle*.

Perhaps the most extensive and clear discussion of the *aesthetic drive*, if we may call it that, takes place on the pages of the text we have been engaging for a while now, *Formulations on the Two Principles of Mental Functioning* (1911). Initially, art is one in the series of compensations for the lost pleasure principle:

> *Art* brings about a reconciliation between the two principles [the pleasure and the reality principles] in a peculiar way. An artist is originally a man who turns away from reality because he cannot come to terms with the renunciation of instinctual satisfaction which it at first demands, and who allows his erotic and ambitious wishes full play in the life of phantasy. He finds the way back to reality, however, from this world of phantasy by making use of special gifts to mould his phantasies into truths of a new kind, which are valued by men as precious reflections of reality. Thus in a certain fashion he actually becomes the hero, the king, the creator, or the favourite he desired to be, without following the long roundabout path of making real alterations in the external world. But he can only achieve this because other men feel the same dissatisfaction as he does with the renunciation demanded by reality, and because that dissatisfaction, which results from the replacement of the pleasure principle by the reality principle, is itself a part of reality. (SE 12: 224)

Years later, Freud repeats this assessment almost word for word, only now making a specific clinical connection between the artist and the neurotic:

> The realm of imagination was seen to be a 'reservation' made during the painful transition from the pleasure principle to the reality principle in order to provide a substitute for instinctual satisfactions which had to be given up in real life. The artist, like the neurotic, had withdrawn from an unsatisfying reality into this world of imagination; but, unlike the neurotic, he knew how to find a way back from it and once more to get a firm foothold in reality. His creations, works of art, were the imaginary satisfaction of unconscious wishes, just as dreams are; and like them they were in the nature of compromises, since they too were forced to avoid any open conflict with the forces of repression. But they differed from the asocial, narcissistic products of dreaming in that they were calculated to arouse sympathetic interest in other people and were able to evoke and to satisfy the same unconscious wishful impulses in them too. Besides this, they made use of the perceptual pleasure of formal beauty as what I have called an 'incentive bonus'. (SE 20: 64–65)

The concept of 'incentive bonus' appears for the first time in Freud's book on *Jokes and Their Relation to the Unconscious* (1905). Art, in this sense, is a form of seduction into fantasy and daydreaming, never an isolated act, confined solely to the creative personality of the artist. If art resonates out into the social world, it is all the more surprising that Freud denies it the power to be the aggregate method of adjustment to cultural reality. All this is underpinned by the apparent powerlessness of psychoanalysis, which "can do nothing towards elucidating the nature of the artistic gift, nor can it explain the means by which the artist works—the artistic technique" (SE 20: 65).

But it is precisely here we should also be a little suspicious, and reevaluate this statement against Freud's own explanation of artistic activity. And the essential point is that art can be quite accurately described as an adjustment technique that mitigates between the broad daylight of the external reality and the murky, dark caverns of our desires, wishes, and drives. The way it accomplishes and carries out this task is not only through the world of fantasy or wishful thinking but also, much more tangibly, in the simulacrum of play. This is how Freud answers the call of discovering ourselves in the artistic activity of creative writing:

> Should we not look for the first traces of imaginative activity as early as in childhood? The child's best-loved and most intense occupation is with his play of games. Might we not say that every child at play behaves like a creative writer, in that he creates a world of his own, or, rather, re-arranges the things of his world in a new way which pleases him? It would be wrong to think he does not take that world seriously; on the contrary, he takes his play very seriously and he expends large amounts of emotion on it. The opposite of play is not what is serious

> but what is real . . . The creative writer does the same as the child at play. He creates a world of phantasy which he takes very seriously—that is, which he invests with large amounts of emotion—while separating it sharply from reality. (SE 9: 143–144)

Given the fundamental importance of childhood to psychoanalysis, the connection to art, as it is given earlier, is more than just tenuous. Freud himself, from the beginning, seems to be engaged in this sort of activity. As early as 1895 in *Studies on Hysteria* he opens the discussion on one of the five hysterical cases in the book (Fräulein Elisabeth von R.) in this manner:

> I have not always been a psychotherapist. Like other neuro-pathologists, I was trained to employ local diagnoses and electro-prognosis, and it still strikes me myself as strange that the case histories I write should read like short stories [*Novellen*] and that, as one might say, they lack the serious stamp of science. (SE 2: 160)

The two modes of play and reality and one's ability to move freely between them (like the artist, or the creative writer does) are indeed the test of the difference between a psychotic stance that confounds the two into *one* and what we could call (for lack of a better word) a condition of normality, which constructs for itself two well-integrated but separate spaces—two *within* the one that produces legibility in the external world like an extra sense organ. Perhaps an 'artificial ear', which listens to its own play, makes things resonate in all seriousness, but is also attuned to the limit as the precise distance between reality and fantasy. And as Freud further points out,

> Language has preserved this relationship between children's play and poetic creation. It gives [in German] the name of '*Spiel*' ['play'] to those forms of imaginative writing which require to be linked to tangible objects and which are capable of representation. It speaks of '*Lustspiel*' or '*Trauerspiel*' ['comedy' or 'tragedy': literally, 'pleasure play' or 'mourning play'] and describes those who carry out the representation as '*Schauspieler*'[14] ['players': literally 'show-players']. (SE 9: 144)

In this manner, we have identified something that seems to be the latent operative principle of the psychoanalytic theory itself, its *aesthetic drive* that reaches out into the space of the real via the mechanism of play, like an amplifier whose technical intricacies connect the internal specifications of its mechanism to the world and its materiality. This connection is found precisely through the output of its particular message, whether that output is a child's game, a piece of creative writing, a painting, a musical/poetic sound, or finally the individually idiosyncratic expression of the modern everyman's anonymous psychology. This is, in the end, what the ego is burdened with—the responsibility to support itself in all seriousness at play with reality.[15] This is the only way that it can acquire enough maneuverability and flexibility to serve as the interpretive/analytical organ that mediates among the three information streams, the three internal voices threatening constantly to overload it.

The ego thus becomes the *dramaturgical self*—the skillfully and artfully trained player. It is literally a *Schauspieler*, an actor, in the sense that it stands in for the id, as its replacement. Almost immediately it also becomes a space of renewal into which new impulses from the id must enter, usually in differently repressed forms. Because it shows itself in this way and through such showing brings the world into relief, the ego makes a weighty matter of the theater of display, although when calibrated correctly, it knows when it is time for the curtain call. The neurotic, in his sensuality, possesses exactly the same store of display-notions but forgets the direction and, even more drastically, forgets himself in the momentum of the staging-gesture, thus fusing reality and playing together, ultimately losing touch with both. All this, in a very condensed sense, is encapsulated in Freud's cryptic phrase at the end of *The Dissection of the Psychical Personality*: "*Wo es war, soll ich warden*" (where the id was, there ego shall be), and it is worth considering the sentences that immediately precede this famous expression, because they relate to the overall effect, the end result that psychoanalysis itself desires, hopes for, and attempts to effectuate:

> It is easy to imagine . . . that certain mystical practices may succeed in upsetting the normal relations between the different regions of the mind, so that, for instance, perception may be able to grasp happenings in the depths of the ego and in the id which were otherwise inaccessible to it. It may be admitted that the therapeutic efforts of psychoanalysis have chosen a similar line of approach. Its intention is, indeed, to strengthen the ego, to make it more independent of the super-ego, to widen its field of perception and enlarge its organization, so that it can appropriate fresh portions of the id. Where the id was, there ego shall be. It is a work of culture. (SE 22: 80)

It is only after carefully considering and emphasizing the division between reality and pleasure, and the role of the ego in the formation of the self, that we can start comprehending the otherwise quite dense and somewhat enigmatic formulations that appear in another of Freud's seminal texts, *The Ego and the Id* (1923). Whereas it is always easy to conclude that the fundamental premise of psychoanalysis is the division between what is conscious and what is unconscious, all the more so because Freud stresses this point on many occasions himself, including at the beginning of the aforementioned text (only to withdraw from such a division when the structural implications of his theory are at stake), this conceptual cliché leaves us empty-handed. Unreflective references to the unconscious distort the shape of what is actually at stake in psychoanalysis. In fact, in a counterintuitive and seemingly self-contradictory gesture, it would be more accurate to say that psychoanalysis is a psychology of consciousness *par excellence*, not only because it is an act of constant and unceasing analysis but especially because the only meeting place where the exchange between psychic energies can take place, and onto whose confines, if the analytical adjustment is to be accomplished, the unconscious information must be impressed, is the ego.

Being thus resigned to the condition of meandering analytical judgments, the ego not only performs a series of cohesive reconciliations with reality through the methodology of listening to the world but also replays the echo of everything that deflects back onto its surface, including the vibrations of inorganic matter. This is how it inscribes the gesture of the ultimate loss of its physical consistency (losing it in effigy, with every psychic readjustment) into the adhesiveness of its psychological awareness. Ultimately, the ego's hard-won independence via readjustment to and reconciliation with all the forces that work against it, the friction that wears it down and desires to rip it apart, is nothing more than simply an aesthetic effect. This might at first strike us as very banal. But aesthetics contains all the theoretical gravity and precision of what Freud already proposes under the auspices of play and its seriousness. This is why art and aesthetics are so seductive. They inspire the moment of finding oneself in the instant of separation, not only from our own ego but also from the 'thing' that now faces us as a foreign territory, in which we must nonetheless identify something, whether it is pleasure or ultimately ourselves:

> the essential *ars poetica* lies in the technique of overcoming the feeling of repulsion in us which is undoubtedly connected with the barriers that rise between each single ego and the others . . . The writer softens the character of his egoistic daydreams by altering and disguising it, and he bribes us by the purely formal—that is, aesthetic—yield of pleasure which he offers us in the presentation of his phantasies. We give the name of an *incentive bonus*, or a *fore-pleasure*, to a yield of pleasure such us this . . . In my opinion, all the aesthetic pleasure which a creative writer affords us has the character of a fore-pleasure of this kind, and our actual enjoyment of an imaginative work proceeds from a liberation of tensions in our minds. (SE 9: 153)

It is precisely the task of the analytic technique to do the same, to find a way of liberating tensions by setting the patient into the mode of play in which he can find himself again. This is a play that not only goes back to one's childhood and its displaced memories or the ancient drama of the Oedipus complex but also something that must be performed currently, whose stage is the presence of unmitigated and nonnegotiable contemporaneous circumstances that must be confronted. We are thus presented with psychoanalysis as a special space that deals with the many barriers arising between each single ego, as Freud says earlier, and the external reality in general; a technique of resolutions that exchanges the overcompensating plasticity of symptoms for the much quieter detachment from reality, that reassures the ego while still satisfying the "ear of conscience", as we have already heard Freud say (SE 19: 57).

This recalibration is accomplished by making subtle adjustments in perception, in the way one recognizes things, the way one apprehends, evaluates, and makes judgments of discrimination among impulses and microrealities. Such aesthetics is already far removed from its standard philosophical embodiment as judgments concerning the formal concept of the beautiful. Rather, it 'staples' together this seemingly unbridgeable distance between reality and what is not covered by its 'blanket statement'. In this way, the ego feels firmly *implanted* in reality that envelops it. An expression of this kind of aesthetics, as we will see, finds its way into the theory of *Einfühlung* (empathy), which constitutes one of the most interesting inventions around the turn of the century.

Freud grants such a central position to the conscious ego, but only cautiously: "psychoanalysis cannot situate the essence of the psychical in consciousness, but is obliged to regard consciousness as a quality of the psychical, which may be present in addition to other qualities or may be absent" (SE 19: 13). The conscious ego is not a pure substance but an effect of perpetual *deflections* among the three forces of id, the superego, and the external reality. These deflections and their echoes as the elementary fabric of individual psychology are the most enduring and important contributions of the psychoanalytic paradigm to modern cultural identity—not the unconscious, which has its own long philosophical history and to which very little can be added, apart perhaps from the renewed acknowledgment of its mysterious presence. And this, in the end, is also Freud's assessment. It is true that

> We have found—that is, we have been obliged to assume—that very powerful mental processes or ideas exist (and here a quantitative or *economic* factor comes into question for the first time) which can produce all the effects in mental life that ordinary ideas do (including effects that can in their turn become conscious as ideas), though they themselves do not become conscious. (SE 19: 14)

But this leads to a series of complications that eventually make the concept of the unconscious as such unworkable. First, Freud distinguishes between two instances of the unconscious, one descriptive and the other dynamic. Descriptive unconscious is but a mere sign, a heard resonance of the dynamic unconscious that attempts to capture the functionality of the entire process. This leads to the realization that whether we see something as conscious or unconscious depends on the point of orientation, or, as Freud says, "the distinction between conscious and unconscious is in the last resort a question of perception, which must be answered 'yes' or 'no', and the act of perception itself tells us nothing of the reason why a thing is or is not perceived" (SE 19: 15–16). The standard of proof, to repeat once again, rests with the ego. This is the main reason why the ego becomes such a complicated structure in psychoanalysis, because it is burdened with the entire measure of the world, bringing it within the bounds of perceived control through decision making:

> We have formed the idea that in each individual there is a coherent organization of mental processes; and we call this his *ego*. It is to this ego that consciousness is attached; the ego controls the approaches to motility—that is, to the discharge of excitations into the external world; it is the mental agency which supervises all its own constituent processes . . . From this ego proceed the repressions, too, by means of which it is sought to exclude certain trends in the mind not merely from consciousness but also from other forms of effectiveness and activity. In analysis these trends which have been shut out stand in opposition to the ego, and the analysis is faced with the task of removing the resistances which the ego displays against concerning itself with the repressed. (SE 19: 17)

As is well known Freud's language is most elementary, based on everyday signifiers and their weightless anonymity. The ego, or simply the 'I' (*das Ich*), is continually responsive to the namelessness of the 'it' (the Id, or *das Es*)[16] and the violent idealism of the superego, the 'over-I' (*Über-Ich*). It is these components that add themselves together, like three harmonic lines, to make up what Freud refers to as the *organism* of the ego. And the curious thing is that the ego contains the repressed, unconscious forces within itself, thus complicating immensely not only the concept of the unconscious but also its own functionality:

> Since . . . there can be no doubt that . . . resistance emanates from this ego and belongs to it, we find ourselves in an unforeseen situation. We have come upon something in the ego itself which is also unconscious, which behaves exactly like the repressed—that is, which produces powerful effects without itself being conscious and which requires special work before it can be made conscious. From the point of view of analytic practice, the consequence of this discovery is that we land in endless obscurities and difficulties if we keep to our habitual forms of expression and try, for instance, to derive neuroses from a conflict between the conscious and the unconscious. We shall have to substitute for this antithesis another, taken from our insight into the structural conditions of the mind—the antithesis between the *coherent* ego and the repressed which is split off from it. (SE 19: 17, my emphasis)

Such realization sets into motion a series of profound consequences, especially for the concept of the unconscious:

> We recognize that the *Ucs.* [Unconscious] does not coincide with the repressed; it is still true that all that is repressed is *Ucs.*, but not all that is *Ucs.* is repressed. A part of the ego, too—and heaven knows how important a part—may be *Ucs.*, undoubtedly is *Ucs.* And this *Ucs.* belonging to the ego is not latent like the *Pcs.* [Pre-conscious]; for if it were, it could not be activated without becoming *Cs.*[Conscious], and the process of making it conscious would not encounter such great difficulties. When we find ourselves thus confronted by the necessity of postulating a third *Ucs.*, which is not repressed, we must admit that *the characteristic of being unconscious begins to lose significance* for us. It becomes a quality which can have many meanings, a quality which we are unable to make, as we should have hoped to do, the basis of far-reaching and inevitable conclusions. Nevertheless we must beware of ignoring this characteristic, for the property of being conscious or not is in the last resort our one beacon-light in the darkness of depth-psychology. (SE 19: 18, my emphasis)

And thus Freud once again ends at the crossroads of self-contradiction, the result of which is a new definition of the entire psychic apparatus in terms of the ego, with unconscious components. It is a topographical mapping of the mind, where particular attention must be paid to scaling and positioning of each component in relation to the other ones, and in this sense we are speaking here of a psychological *composition* that operates on the basis of a distributive paradigm that is bound to the principle of differentiation as transfer, flow, and displacement between levels of intensity.

Given the conclusion that the contrast between consciousness and unconsciousness leads to irreconcilable ambiguities that stymie any kind of adequately focused view into the functioning of the human mind, any subsequent investigation must be bound to the system of consciousness, in its general and specifically theoretical expressions:

> Now all our knowledge is invariably bound up with consciousness. We can come to know even the *Ucs.* only by making it conscious . . . We already know the point from which we have to start in this connection. We have said that consciousness is the *surface* of the mental apparatus; that is, we have ascribed it as a function to a system which is spatially the first one reached from the external world—and spatially not only in the functional sense but, on this occasion, also in the sense of anatomical dissection. Our investigations too must take this perceiving surface as a starting-point. (SE 19: 19)

Having identified such a starting point, the task now becomes to figure out and describe how anything at all can become conscious.

The crux of the matter comes down to the way things bond with other elements, like tones creating acoustic compounds and thus new substances, in order to arise as conscious perceptions. When the information stream comes from the direction of the external reality, the bonding mechanism is that of word presentations; when it originates internally it manifests itself through memory traces. But the bottom line is that in order to have a psychologically significant effect, the psychic impulses must be transformed into perception. It therefore follows that

> The question, 'How does a thing become conscious?' would thus be more advantageously stated: 'How does a thing become preconscious?' And the answer would be: 'Through becoming connected with the word-presentations corresponding to it'. (SE 19: 20)

The interesting thing is that word presentations are the reconstituted *acoustic resonance* of memory traces:

> Verbal residues are derived primarily from *auditory perceptions*, so that the system *Pcs.* has, as it were, a special sensory source. The visual components of word-presentations are secondary, acquired through reading, and may to begin with be left on one side; so may the motor images of words, which, except with deaf-mutes, play the part of auxiliary indications. In essence a word is after all the mnemic residue of a word that has been *heard*. (SE 19: 20–21, my emphasis)

It is at this point that we encounter the kernel of the psychoanalytic procedure that simply mediates between the external and the internal environments, by attempting to facilitate new paths of connection between them via language as word presentations (the talking cure). This positioning of language, to take a clue from a writer, is, as William Burroughs remarks, "a component of the body like any other [because] words are micro-organisms, living dust",[17] which psychoanalysis reforms and rearranges. And thus the sonority of sound, as the paradigm according to which the turn-of-the-century expressive/existential ventures operate, comes out into the open from the debris of the psychoanalytic method.

The importance of hearing and sound to the construction of perception is emphasized by Freud in the adjoining diagram to the text of *The Ego and the Id*, as well as in this somewhat 'rugged' sentence: "We may add, perhaps, that the ego wears a 'cap of hearing' [*Hörkappe*]—on one side only, as we learn from cerebral anatomy. It might be said to wear it awry" (SE 19: 25). Sound is indispensable to consciousness, which, as it turns out, is a more important structural/psychological element in Freud's psychoanalytic theory than the fame of the unconscious would indicate. It is the principal element of identity's construction, its sustenance as well as stability. This is further accentuated in how Freud confounds images and the unconscious:

> Thinking in pictures is, therefore, only a very incomplete form of becoming conscious. In some way, too, it stands nearer to unconscious

> processes than does thinking in words, and it is unquestionably older than the latter both ontogenetically and phylogenetically. (SE 19: 21)

Now we can also understand the intriguing formulation that consigns the death drive to a soundless, mute condition: "we are driven to conclude that the death instincts are by their very nature mute [*stumm*] and that the clamour of life proceeds for the most part from Eros" (SE 19: 46). But even here it is important to point out that muteness is not total silence. The death drive is still an *acoustic arrangement* whose volume, if life is to proceed, must be turned down but it can never be completely eliminated. Eros or the libido as the life-affirming drives constitutes this very unique part of the id that can never be quite sublimated, repressed, and quieted down—hence the importance of the sexual amortization in the life of an individual, as well as in Freud's analytical-theoretical constructions. But the importance and meaning of the contrast between noise and noiselessness, sound and silence, reach beyond this immediate concern, because for anything to appear at all, for it to become consciousness and thus exhibit itself as part of the ego, it must be amplified through words and their sonorous inflection. Sound, then, is not simply a metaphor, an analogy for how everything works, but *the operational principle*, the notation key under whose auspices reality comes into its own.

Having thus exposed a latent but absolutely crucial theory of sound in psychoanalysis, we may now begin to understand better the positioning of the ego in the entire setup, its activity as a method that translates the internal and the external stimuli into *coherent 'organicism'* "whose supreme goal is to represent organic motifs in all their accidental and transitory momentary appearances of a receptive surface":[18]

> It is easy to see that the ego is that part of the id which has been modified by the direct influence of the external world through the medium of the *Pcpt.-Cs.* [Perception-Consciousness]; in a sense it is an extension of the surface-differentiation. Moreover, the ego seeks to bring the influence of the external world to bear upon the id and its tendencies, and endeavours to substitute the reality principle for the pleasure principle which reigns unrestrictedly in the id . . . The ego represents what may be called reason and common sense, in contrast to the id, which contains the passions. (SE 19: 25)

The relationship between these two surfaces is predominantly aural-auditory, given how its identity can be found only in repackaged acoustic traces (words), their signification carried by their tone, just like Anna O.'s stream of consciousness, whose vernacular forgets the formal linguistic restrictions and simply sounds out its message in a play of showing and vanishing. This type of engagement is a form of reality-testing, similar to the *fort/da* (gone/there) play of Freud's grandson, able to redefine distances and intervallic gaps. The boy comes to terms with his mother's frequent and sudden absences by resorting to the topographically choreographed activity of letting various objects go and then retrieving them.[19] It is a moment of deliberately generated removal that attempts to gain mastery over unexpected absence. This play with absence, in which audition is by default engaged given how sound enters where vision dissipates, proves crucial not only to psychoanalysis but also to the entire *fin-de-siècle* setting, whose rhythm attempts to coordinate a series of crucial existential deficiencies by bringing forth and letting go of phenomena as one tries to imitate the missing quality of experience.

Freud reaches a similar, acoustically inspired conclusion towards the end of *The Ego and the Id*, when he attempts to bring the concept of the super-ego into the fold of the general scheme: "Our tentative answer will be that *it is as impossible for the super-ego as for the ego to disclaim its origin from things heard*; for it is a part of the ego and remains accessible to consciousness by way of these word-presentations (concepts, abstractions)" (SE 19: 52, my emphasis).

Acoustics as theme and procedure is also unmistakably present on the pages of Freud's writings about the intricacies of the analytical technique. For instance, in *Recommendations to Physicians Practicing Psycho-analysis*, Freud uses the following stark metaphor:

> Just as the patient must relate everything that his self-observation can detect, and keep back all the logical and affective objections that seek to induce him to make a selection from among them, so the doctor must put himself in a position to make use of everything he is told for the purposes of interpretation and of recognizing the concealed unconscious material without substituting a censorship of his own for the selection that the patient has forgone. To put it in a formula: he must turn his own unconscious like a receptive organ towards the transmitting unconscious of the patient. He must adjust himself to the patient *as a telephone receiver is adjusted to the transmitting microphone.* Just as the receiver converts back into sound waves the electric oscillations in the telephone line which were set up by sound waves, so the doctor's unconscious is able, from the derivatives of the unconscious which are communicated to him, to reconstruct that unconscious, which has determined the patient's free associations. (SE 12: 115–116, my emphasis)

This procedure amounts to what Freud designates in the same essay as *gleichschwebende Aufmerksamkeit*, or "evenly-suspended attention":

> The technique [of psychoanalysis] is a very simple one. As we shall see, it rejects the use of any special expedient (even that of taking notes). It consists simply in not directing one's notice to anything in particular and in

> maintaining the same 'evenly-suspended attention' (as I have called it) in the face of all that one hears . . . Or, to put it purely in terms of technique: 'He [the analyst] should simply *listen*, and not bother about whether he is keeping anything in mind. (SE 12: 111–112, my emphasis)

In other words, the analyst should simply let the patient's voice play. Experiences that are vocally expelled from the thin surface of unimportant everyday events, along with their static interference, bring up sonic structures from the inner, unconscious depth, thus condensing silent impulses into channels of heard communication that analysis is meant to facilitate. In this manner, psychoanalysis triggers the condition of things playing themselves out as they are repeated, for the repressed possesses the uncanny quality of *Widerholungszwang*, the rhythmicity of the 'compulsion to repeat'—yet another expression of the musical 'algorithm' whose general outline is already seen in the mentioned *fort/da* mechanics of distancing and bringing closer.

Freud's famous consulting room is a space where one needs to learn how to listen, and moreover how to listen to previous listening as a process of developing variation. One is able to do this by ever new, evolving *arrangements* of sound. In the history of Western music, as Peter Szendy observers, arrangement played this function of a space where one's listening as reworking of someone else's composition, very often quite drastic, was signed and transmitted.[20] The consulting room, in which analysis takes place, begins the process of acoustic retrieval through a specific arrangement of positions:

> I hold on to the plan of getting the patient to lie on a sofa while I sit behind him out of sight. This arrangement has a historical basis; it is the remnant of the hypnotic method out of which psycho-analysis was evolved. But it deserves to be maintained for many reasons. The first one is a personal motive . . . I cannot put up with being stared at by other people for eight hours a day (or more). Since, while I am listening to the patient, I, too give myself over to the current of my unconscious thought, I do not wish my expression of face to give the patient material for interpretation or to influence him in what he tells me. The patient usually regards being made to adopt this position as a hardship and rebels against it, especially if the instinct for looking (scopohilia) plays an important part in his neurosis. I insist on this procedure, however, for its purpose and result are to prevent the transference from mingling with the patient's associations imperceptibly, to isolate the transference and to allow it to come forward in due course, sharply defined as a resistance. (SE 12: 133–134)

This is how the process that configures symptoms, cases, and disorders begins: by closing the eyes

5 Echoes

Echoes are the acoustic copies of the now uprooted voice. They are counteracting the condition of being "buried within ourselves [where] we send out signals, gestures, and sounds indefinitely and uselessly. No one listens to anyone else. Everyone speaks; no one hears; direct or reciprocal communication is blocked."[1] By 1938, Theodor Adorno diagnosed this echoless condition as the regression of listening—an unmistakable "reduction of people to silence, the dying out of speech as expression, the inability to communicate at all." This loss of the *resonant ego* "inhabits the pockets of silence that develop between people moulded by anxiety, work and undemanding docility"[2]—precisely the symptoms that the psychoanalytic recording studio was set up to neutralize.

But what exactly happens there? One of the last things Freud ever wrote on the 'nature' of the psychic apparatus, dated August 22, 1938, and published posthumously, reconfirms what we have already uncovered in a slightly different inflection: "Space may be the projection of the extension [*Ausdehnung*] of the psychical apparatus. No other derivation is probable. Instead of Kant's *a priori* determinants of our psychical apparatus. Psyche is extended; knows nothing about it [*Psyche ist ausgedehnt, weiss nichts davon*]" (SE 23: 300). The analytic session is precisely the setting where the psyche learns what it knows nothing about. It hears itself in a series of encounters, blindly running into things such as objects and memories, actively constructing complex networks of perception. Kant's *a priori* space is inconsequential because it is outside consciousness. One is simply thrown into it. We get *in touch* with space only by inhabiting it through various forms of confrontation. But this system is unstable. Consciousness is a quality that appears and disappears, comes in and goes out, protrudes and recedes. It is full of breaks and gaps that need to be filled in:

> This reminds us that consciousness is in general a highly fugitive state. The state of being conscious is at best momentary. If our perceptions do not confirm this, the contradiction is only an apparent one; it is explained by the fact that the stimuli which lead to perception may

> persist for considerable periods, so that meanwhile the perception of them may be repeated. (SE 23: 159)

In other words, consciousness is a composite not a monolithic texture, analogous to Schoenberg's approach to composition based on repetition of acoustic shapes and patterns, which, despite their atomization, condense into a unified dimension of sound: "The main advantage of this method of composing with twelve tones is its unifying effect", wrote Schoenberg late in his creative life.[3] But throughout he worked on isolating a sharply delineated musical consciousness that had the ability to shape distances between notes, instead of receiving them ready-made from the classical theory of harmony. And, more often than not, finding consciousness meant that one had to invent what should not exist, as was the case already in 1899, when a musical society refused to perform Schoenberg's sextet *Verklärte Nacht* because of the inverted ninth chord, which, as everyone knew, simply does not exist. And the nonexistent is distanced, quarantined from possible encounters—it is confined to silence.

Among the many equivalent and progressively narrower formulations of the psychoanalytic practice, we find one that sets the concept of distance at its center: "For the work of analysis proceeds best if the patient's pathogenic experiences belong to the past, so that his ego can stand at a distance from them" (SE 23: 232).

The purpose, as always, is to bring the pressure of the internal impulsive life into reconciliation with the reality at large. Here, we must remind ourselves that Freud imparts this kind of ability for coherent synthesis to the artist, but he also mentions the concept in relation to the patient, who is to undergo the procedure that synthesizes auditory traces of everything that is spoken during analysis. In the end, this can be accomplished only by the unprecedented flexibility of the ego:

> Not every neurotic has a high talent for sublimation; one can assume of many of them that they would not have fallen ill at all if they had possessed the art of sublimating their instincts . . . It must further be borne in mind that many people fall ill precisely from an attempt to sublimate their instincts beyond the degree permitted by their organization and that in those who have a capacity for sublimation the process usually takes place of itself as soon as their inhibitions have been overcome by analysis. (SE 12: 119)

It is precisely the artist who displays such a heightened ability for sublimation, and in that sense the difference between a 'normal' state and a disorder is only one of degree in the ego's ability to flex its own structure in the direction of the id, and then be strong enough to return from such a journey back to reality. What distinguishes the ego from other psychic enclosures that analysis is specifically designed to redress in its deficiency is *plasticity.*

On this basis, Freud differentiates among different types of identity, more or less prone towards the analytic adjustment as a function of the plastic ability of the ego. This is how:

> We come across people, for instance, to whom we should be inclined to attribute a special 'adhesiveness of the libido'. The processes which the treatment sets in motion in them are so much slower than in other people because, apparently, they cannot make up their minds to detach libidinal cathexes [*Besetzungen*—i.e., occupations] from one object and displace them on to another . . . One meets with the opposite type of person too, in whom the libido seems particularly mobile; it enters readily upon the new cathexes suggested by analysis, abandoning its former ones in exchange for them. The difference between the two types is comparable to the one felt by a sculptor [*der bildende Künstler*], according to whether he works in hard stone or soft clay [*der Ton*]. Unfortunately, in this second type the results of analysis often turn out to be very impermanent: the new cathexes are soon given up once more, and we have an impression not of having worked in clay, but of having written on water . . . In another group of cases we are surprised by an attitude in our patients which can only be put down to a *depletion of the plasticity*, the capacity for change and further development, which we should ordinarily expect. We are, it is true, prepared to find in analysis a certain amount of psychic inertia . . . But with the patients I here have in mind, all the mental processes, relationships and distributions of force are unchangeable, fixed and rigid. (SE 23: 241–242, my emphasis)

This is more than just an illustrative metaphor but the paradigm that underpins psychoanalytic theory—itself an offprint of the historical model of cultural identity circa 1900.

The English translation seems to overadjust the text. Freud refers to artists (*bildende Künstler*), not specifically to the sculptor (*der Bildhauer*), and the soft clay (*der Ton*) that is set in opposition to stone suggests the entire, elastic spectrum of resonance—enclosed in its own etymological chain. The tone, from the Greek *tonos*, tension, *teinein*, to stretch, is all flex and vibration. And from the physiological perspective, this acoustic identity found in the toned tissues, their firmness and elasticity, is an indication of the state of health.

Freud is after a type of alignment absolutely necessary for the ego to be evocative of itself and responsive to analysis. Such responsiveness depends on the bonding power of the plastic element, whose consistency is more or less receptive to structural manipulation. The end result must be a form that supports the reconstructive project, organizing, harmonizing, and releasing cognitive dissonance, as dictated by specific circumstances. This is how the cultural identity at the beginning of the last century acquires the quality of an *ontological tonality*[4]—something that is not an affirmation but the incessant questioning of spatial dimensions: "What does it mean for a being to be immersed entirely in listening, formed by listening or in listening, listening with all his being?"[5] What it means is *dissipation* triggered by "dissonance which rejects belief in the illusion of the existing harmony."[6]

Such a dissonance enables a consciousness that, despite being "the one light which illuminates our path", is still only "an incomplete and broken chain of phenomena" (SE 23: 286). This is why, as Freud continually insists, consciousness is ever only a sporadic quality of the entire mental apparatus, surrounded by the unceasing background noise of everything outside its momentary control. This is the style of thinking permeating the entire *fin de siècle*—thinking through mutually overlapping layers of art, aesthetics, philosophy, psychology, and culture in an attempt to merge consciousness with the always escaping point of contact, carried away by the continually dislodged 'truth'. This fugitive progression and its expectation are enclosed in the commotion of the ego's own bodily presence, where

> Pain, too, seems to play a part in the process, and the way in which we gain new knowledge of our organs during painful illnesses is perhaps a model of the way by which in general we arrive at the idea of our body. (SE 19: 25–26)

The body, akin to a percussive instrument, refines its materiality into textures of consciousness, performing a series of analyses on its own constitution. Freud expands on this theme, by including a footnote to the English translation of the *Ego and the Id*, where we have already encountered the somewhat cryptic formulation of the ego's corporality. The footnote, which does not appear in the German version of the text, reads,

> The ego is ultimately derived from bodily sensations, chiefly from those springing from the surface of the body. It may thus be regarded as a mental projection of the surface of the body. (SE 19: 26)

In essence, the ego is a transplanted, diffused, and amplified materiality. It is something like the body's virtual organ, a phantom limb, an effigy, invested with very similar functionality—that of holding the entire mass of processes, parts, and systems together, in a state of momentary equilibrium. And "suppose it becomes an organism, a homunculus or a robot, and possesses some of the spontaneity of a vision; it remains yet another thing to organize this form so that it becomes a comprehensible message 'to whom it may concern'", Freud could have said, if it was not Schoenberg speaking about the type of generative process that informs his method of composition.[7] 'To whom it may concern' is also the ultimate addressee of psychoanalysis. That the message concerns the nameless everyone expresses the general cultural identity of that time.

The ego then, as extension of the body, acts like a resonant membrane through which every perception must pass and leave a mark, so that it can be registered as a trace on the plane of reality. All these marks, when added together, compose a psychology. Before the fully developed, modern psychic apparatus based on abstract thinking took hold, the body was standing over and against the world very much on its own. The fact that a large part of the ego remains under the tension of resistance to things that threaten its existence is yet another echo of the body, which, after all, specializes in resisting exercises if only through its mechanical setup. In fact, the ego cannot exist without the readiness and praxis of muscle flexing. And, at one point in this constantly spiraling discussion, Freud lets the two forms, the physical and the psychological, merge:

> As a frontier creature, the ego tries to mediate between the world and the id, to make the id *pliable* to the world and, by means of its muscular activity, to make the world fall in with the wishes of the id. In point of fact it behaves like the physician during an analytic practice: it offers itself, with the attention it pays to the real world, as a libidinal object to the id, and aims at attaching the id's libido to itself. (SE 19: 56, my emphasis)

Pliability, attachment, and mobility are acoustic qualities. This is the only way one finds "an approach to . . . the form or structure of *self* as such, that is to say, to the form, structure, and movement of an infinite referral, since it refers to something (itself) that is nothing outside of the referral."[8] We may also describe such a referral as the echo of one's material presence because "when one is listening, one is on the lookout for a subject, something . . . at once the same as and other than itself, one in the echo of the other, and *this echo is like the very sound of its sense*."[9]

It is only art that possesses the existential know-how and the practical experience that allow one to remain in this circulatory system of echoes—an orientation that breaches the surrounding environment surprisingly effectively. This kind of solution might seem to be very banal, even superficial, but for the ego it means the world. That world is structurally supported by the aesthetics of disappearance. Identities circa 1900 are caught in its vortex, in the perspective of their own fleeting centrality, and their own paradoxically desired intransigence.

***

The ego is a hybrid formation, able to populate and thus placate the unwieldy world with its own constructions. Problems occur when it believes too much in the 'truth' of its own self-image (neurosis) or disbelieves its own presence, tuning out of the world completely and withdrawing into itself (psychosis). The balance between these two positions depends on how effective the ego's acoustic imagination is, given how the voice and its plastic logic provide the most direct path of engagement with phenomena. We should remind ourselves that not only the ego but also the superego are based on things heard (SE 19: 52), and that the more primitive, image-oriented unconscious can be apprehended only by amplification that reformulates it, as much as possible, into sound-images. We can arrive at the unconscious parameters of presence only by first translating it into something conscious (SE 14: 166). This is how word presentations are linked to the object presentations—through sound-images (SE 14: 214). Moreover, this acoustic imagination is a neutral, plastic energy, full of untapped potential, although Freud is not quite certain as to its location and identity:

> We have reckoned as though there existed in the mind—whether in the ego or in the id—a displaceable energy, which, neutral in itself, can be added to a qualitatively differentiated erotic or destructive impulse, and augment its total cathexis [occupation/*Besetzung*]. Without assuming the existence of a displaceable energy of this kind we can make no headway. The only question is where it comes from, what it belongs to, and what it signifies. (SE 19: 44)

Such displaceable energy that allows for new forms of identity to be put together by rearranging old components directs identity towards a fundamentally 'artistic individuality' that looks for its own style of existence. By the time the twentieth century painfully arrives at its middle age, this formula becomes even more obvious, allowing Marshall McLuhan in 1960 to simply proclaim that "we must all become creative artists in order to cope with even the banalities of daily life."[10]

We can perhaps understand better now how and why the *fin-de-siècle* culture, as it tried to establish its own historical parameters, might have wanted to turn everything slightly askew by applying a new existential algorithm—using its cultural identity to continually readjust the shape of its life-world. Various phenomena gave up their traditional identity, only to become newly engineered *organs* and reinserted immediately into the flow of things. The entire Freudian psychic apparatus is an expression of this type of organic engineering, as is Schoenberg's acoustic system. According to the logic of the *Augenmusik*, such a play of surface and space becomes a 'musicology'—the logic of what Ernst Bloch in 1918 described as *Hellhören* or 'visionary hearing'. Let us look for it in some other period-specific spaces.

PLASTIC MANIPULATIONS

"With a shaving knife cut the pith of a sunflower into slices. Observe the miniature that appears in white on gold, and you will see a man's image," wrote August Strindberg in his 1897 essay on the sunflower.[11] This formulation expresses the type of aesthetic inscriptions the *fin de siècle* revels in, where definitions of materiality are extended into avenues of improbability, uncomfortable conviction, and in the end, especially from the perspective of future scientific developments, strangely amusing correctness.[12]

Strindberg was a modern-day alchemist who pursued experimentation abstractly and empirically. He found meaning in the most startling spaces, like in the structure of the sunflower, where the miniature panorama of the human form and the echo of cultural forces are heard. A series of such plastic juxtapositions, most famously expressed in van Gogh's obsession with sunflowers, was one of the triggers that launched expressionism.[13] Strindberg is a psychological designer, filling the gaps in consciousness, recomposing its momentary, in-and-out presence. At the end of his life, Freud assessed the psychoanalytic project in the same way, observing that "we fill in what is omitted by making plausible inferences and translating it into conscious material." It is only "in this way we construct, as it were, a sequence of conscious events complementary to the unconscious psychical processes" (SE 23: 159). Schoenberg, too, will propose that the sequence of twelve tones he, at one point, makes into the spine of his acoustic procedure is but an expression of an unconscious and otherwise mute artistic impulse, not merely an elaboration of an established technique of how musical notes are to be connected and squeezed into musical forms. This sort of stance is just another echo of the psychology of expectation that attempts to suture the gaps between the numbness of merely organic substance and human life, elevating them to the position of new, hybrid, and resequenced identity.

Hence, just as we find humanity in the sunflower, we can also glimpse its floral structure in the physical, cellular makeup of the human body, especially the eye:

> If one makes a cross-section through the eye's cornea, there is a sunflower. When packed with seeds, the sunflower's thalamus presents a honeycomb with cells, with the seeds resembling bee larvae; but in its entirety it also resembles an insect's eye. In a rough sketch the hermaphrodite inner flowers of the sunflower resemble the rods of the retina.[14]

This existential equation, where Analogies = Correspondences = Harmonies (Strindberg's subtitle to his essay on the sunflower), permeates the *fin de siècle*. It is also the formula of psychoanalytic practice throughout, because "the equation 'perception = reality (external world)' no longer holds" (SE 23: 162). And Strindberg ends with the following formulation: "We have seen similarities everywhere, for the simple reason that similarities and correspondences are to be found everywhere, and those who say that they believe in the unity of matter—and the spirit—agree with us! Isn't that so?"[15] Analogies become documents or empirical indicators of presence.

If correspondences are to be found everywhere, then Schoenberg's music is the prime exponent of their testing ground, of finding them precisely there, where they are not supposed to be present. Sound, in his hands, becomes the startling medium of overlapping, acoustic analogies. In other words, Schoenberg creates a sonority whose resonance continuously strips and resurfaces acoustic perception. This procedure is very similar to Freud's analytic procedure that extracts misalignments, using them as new construction materials.

It is a well-known fact that Schoenberg was very much influenced by Strindberg. His library contained the writer's complete works, and at one point Schoenberg seriously considered setting one of his plays to music, a project that in the end had to be abandoned, given Strindberg's untimely death.[16] One could speculate about the attraction Strindberg's idiosyncratic philosophy of experimental empiricism held. The most obvious explanation would simply be its condensing power, concentrating a multitude of forces in the most insignificant phenomena—something that matched Schoenberg's own condensation of sound in the miniatures of single tones. Music is not simply a technical issue of artistic production. For Schoenberg, it represents a universe under threat and in constant need of recovery through the most precise, momentary elaboration of its acoustic identity. Schoenberg's biographer, H. H. Stuckenschmidt, emphasized the impact of Strindberg on the creative development of the composer. In 1909 "on the blank side of a letter from Hermann Bahr, Schoenberg doodled, as he often did, a shape which represented nothing and next to it in large letters the two names AUGUST STRINDBERG—ARNOLD SCHOENBERG."[17] He also points to the similar 'spiritual' development of the two artists whose names resonate together in an almost uncanny homophony. The impact of Strindberg's *Jacob Wrestling* on Schoenberg's later opera *Jacob's Ladder* is unmistakable.[18]

***

Vintage psychology circa 1900 works on the premise of de-ranged and rescaled sensations. In other words, if the one-to-one correspondence between perception and space no longer holds, one needs to find a bypass procedure where connections could still be located.

Strindberg explores this precise issue in another one of his short essays, written in 1894/1895 under the title of 'Sensations Détraquées' and published in the Parisian literary magazine *Le Figaro littéraire*. The title 'Sensations Détraquées' carries several connotations. The phrase 'deranged sensations' seems to point to a state on the precipice of madness, of being out of touch with the world, distorting the grace of its welcoming stability. This standard signification, as we learn from Strindberg's essay and the meaning that 'sensations détraquées' carried in the milieu of *fin-de-siècle* Paris, is completely turned around. In the literary circles at the time, 'deranged sensations' was a label specifying the condition of extreme creative sensitivity and intellectual intensity—a characteristic that became the core of any artistic personality, one to be possessed not only on those irregular occasions when the creative frenzy takes over but all the time, as part of one's basic constitution and character. For Strindberg, this condition of distraction and derangement that sees the world differently from what is available through the standard sensory apparatus takes on another, even more significant meaning. He considers this the only state that can ensure the future—its artistic and existential survival. In 1894, Strindberg wrote to a friend,

> Some time ago I sent you an article from *Le Figaro*: "Sensations détraquées", in which I tried, in a favourable mood I'd worked up, to anticipate the capacities of a future, more highly developed mental life, which we still lack, and which I can only conjure up momentarily—before falling back, exhausted by the effort, into my old frame of mind.[19]

The opening section of the essay is a "symbolic, détraquistic compromise between science, poetry, and madness."[20] But it is the anticipation of future psychological capacities that is especially interesting here, especially because the future Strindberg identifies is not really that far off, and in fact already present in its nascent form, including through his own personality. This is the future that in about a decade will be firmly connected to and identified with phenomena such as psychoanalysis and Schoenberg's practice of sound, among others.

The text on 'deranged sensations' describes a train journey from Austria to France, and then a sojourn at a friend's villa in Versailles. Strindberg experiences a sensory vertigo brought on by the intensity of encounters, where distorted topographies and strange reorchestrations of familiar places take hold. It is a head-on confrontation with the meaning of a time in limbo—a misalignment between movement and confinement, the eye and the ear:

> For all forty-eight hours I have sat like a prisoner in a railway carriage, forced to breath in the carbon dioxide and nitrogen of people I do not know. To begin with I simply loathed them, for they disturbed me, these creatures who made me learn their features and compelled me, forcibly, to listen to their conversation, which agitated my mind . . . And from the bottom of my heart I cursed my companions, shut up in the same box as I was. But when tiredness got the better of them, and they fell silent, their faces assumed such sorrowful expressions that I ended up feeling sorry for them. Uprooted from their normal sphere, from their loved ones and from their habits, they filled me with compassion. A general feeling of discomfort settled upon this mean and trying intercourse in the railway compartment . . . and when night had fallen, when these poor people had gone to asleep [sic], with their unwashed hands pressed together upon their stomachs and their pale,

> sweat-drenched faces bent down over their breasts, our compartment resembled a battlefield strewn with corpses and the scattered remains of human limbs. Sleep brings no feelings of happiness, and our torture chamber resounds with sighs, sighs from creatures who, after millions of years of civilisation, have relapsed into an animal or savage state, and dream of green pastures, or perhaps also of a good murder, rape or incest![21]

The violence of the passage, the graphic nature of the mangled anatomy of bodies and spaces, is not at all surprising, given what we have already learned about the *fin de siècle* and its profile. It not only anticipates the great 'theater of cruelty' (Artaud), the social version of an anatomy lesson in the form of the two world wars, but also once again accentuates the type of logic that modernity hides within its confines and according to which it *expects* something of itself—anything in fact, even cruelty, as long as the act of expectation is maintained. By 1932, Antonin Artaud was able to write that "without an element of cruelty at the foundations of every spectacle, the theater is not possible. In the state of degeneracy, in which we live, it is through the skin that metaphysics will be made to reenter our minds."[22]

It is a lesson in dissection and fragmentation, an aesthetics of disappearance that makes *inhabited* ruins out of clearly demarcated enclosures, whether these are material objects, a matter of knowledge, or physical bodies. This is the only way, given the erosion of religion, that the necessity of being expectant of something at all can be maintained. Now almost instinctively, given that the expectation for a promised fulfillment cannot be counted on, the future and its uncertain potential are propelled forward by the unsettled plasticity of the immediate present. And the qualities deemed undesirable are dissolved in a shifting vacuum of events and circumstances that evacuate their contents, practically on command. The present is filled with abrasive substances that can once again transfuse life into the skin and bones.

After the strenuous train journey, Strindberg finally arrives at Versailles and feels emotionally liberated, splicing the present and the past into a new collection of facts. It is an exercise in draining transpositions, where closeness and remoteness are felt together and the nonchalance of disinterestedness predominates: "Even the most deeply rooted of my emotional ties has relinquished its grip and been replaced by a liberating indifference."[23] It is only such indifference that can define the contours of a fresh encounter with places already familiar:

> With my sight sharpened by anger I measure the distance and fancy I can distinguish some busts on the central section of the building . . . seen from the front . . . I take out my map of Versailles, calculate the distance once again, and find that I have only 500 meters to go to reach the chateau, since the entire length of the avenue does not exceed 1000

> meters. Surprised by this simple fact, I explain the whole business to myself as follows: 'the perspective changes as I stride along; at the same time, the optic angle grows larger, and this infernal play of invisible lines confuses my mind, where the irradiating rays of the enchanted castle imprint themselves.[24]

The experience becomes one of overburdened perspective and rationality, where the linear approach to the *space as a limit* is given up in favor of *space as environment*. This is how in 1934 Henri Focillon explained the divergent identity of such spaces:

> Space as limit not only moderates the proliferation of relief, the excesses of projection, the disorder of volumes (which it tends to block into a single mass), but also strongly affects the modelling. It restrains its undulations and disturbances; it suggests the modelling itself merely by an accent, by a slight movement that does not break the continuity of planes . . . However, space as environment, exactly as it delights in the scattering of volumes, in the interplay of voids, in sudden and unexpected perforations, so does it, in the modelling, welcome those multiple, tumbled planes that rend the light asunder.[25]

In other words, the limit is not an external point of flight but an internal dimension that generates reality out of itself, like the body devouring and becoming its own projected point of constantly delayed decomposition, which we know as the ego. Using the conceptual vocabulary still to be established, it is precisely ruins that operate according to such environmental spaces where dimensions are rendered asunder and thrown back together again.

If extensive spaces or spaces as environment break apart the homogeneity and continuity of planes and surfaces, then one should also easily surmise that the entire mass of sound, the way we find it in Schoenberg's compositions, in their entire range, is all about the scattering of volumes and the interplay of voids, of sudden perforations that stretch and split apart the homogeneity of the tonal space and its limits. The Freudian psychic apparatus, its entire scheme, is also such an environmental space—a membrane that resonates through forms and formulas of encounters between the mind and whatever surrounds it, liberating things out of their old enclosures and thus formulating a psychology that is space as much as it is a theory.

Strindberg became a virtuoso in perpetuating, performing, and composing the tensions that lead to such an elementary imbalance, a carnage of perceptions that extends from the railway carriage to the immaculate space of the Versailles gardens and palace:

> Certainly, I want to reach my goal, the palace, but at the same time I want to defy that superior force. My brain divides into two parts and

> joins battle with itself, and I expect to see half my body stroll across the Place d'Armes while the other half remains by the post. I seek in vain to connect the two parts of the machine and strive to discover a self which stands above my self.[26]

The transfiguration of space is especially accomplished musically, by piercing the anatomy of volumes that are out of range:

> I become aware that the Cour de Marbre comprises the auditory canal of a huge ear, the auricle of which is formed by the wings of the building. Captivated by this new fantasy and happy to have come upon this bizarre idea, that I am like a flea in a giant's ear, I listen pressed against the wall . . . What a surprise! . . . I can hear! . . . For it is Paris that I can hear murmuring along this chain of hills, which stretches from Courbevoie to Sceaux and spreads out in a semicircle where the principal focus is Versailles, and the auditory canal is the Sèvres vally.[27]

The experience, whose confusion forces Strindberg to ask if he is 'out of kilter', because he was born when people had six-volume novels to read, or if he is 'changing skin' due to the nerves that are undergoing evolutionary crisis, also repeats itself in the most intimate of spaces, the bedroom where, in a restless mood, Strindberg takes his guitar and retunes his nerves:

> It is a habit of mine to tune my soul and the instrument to each other in this way, and when I feel depressed, I raise my soul tone by tone while tuning the pegs of the guitar. Today my nerves are in the key of D minor.[28]

Just a few years later, Schoenberg will retune his own nerves and those of his contemporaries, eventually without a key at all. But before that takes place his favorite mode of interaction with sound is also D minor.[29]

By the time Strindberg had his fill of nervousness, the nerves became the general, psychological condition of modern life. Their frayed textures, pictured in diagrams as if the human body was overgrown from the inside by nature's map of roots and branches, seemingly throw identity down the evolutionary scale, closer to the condition of plant life. This is where environmental sensitivity, fragility, and instability combine and threaten to wipe out entire habitats—a stance clearly in contradiction of the evolution-inspired adaptability and resilience. Strindberg, almost instinctively, falls under the spell of nervousness in its plant-like character:

> I contemplate a gigantic pine in front of me . . . Like an animal which has come into being through crossbreeding between a polyp, an insect, a snake and a fish, the pine towers up into the heights. Its elongated body, all covered in scales, encloses the circulatory and lymphatic

> system . . . The pine sinks its open belly down into the earth, its outer and inner intestines secreting digestive juices over its food before it sucks it up, just as the fly does . . . My pine is a living being, a large animal that eats, digests, grows and loves . . . He loves this pine. Hence he has nerves! Where? . . . They exist everywhere! . . . Enough of this! He feels, my pine.[30]

This encounter and its experimental character, a distinction that is somewhat of a misnomer because for Strindberg all phenomena worthy of the name are experiments in progress, are a clear instance of space as environment—a space that had lost its parameters. Its topography was now governed by chance or at least the kind of expectation that chance inadvertently brings with it.

Chance and the liberating, unpredictable potential it hides are the theme of another short essay. In *The New Arts! or The Role of Chance in Artistic Creation* (1894),[31] as the title already suggests, Strindberg uses the logic of decay, musing about coherence and transgressions that erase symmetry. At stake is the singularity of perception, illustrated with a musical example:

> They say that the Malays make holes in the bamboo stems that grow in their forests. When the wind comes the savages lie on the ground and listen to symphonies performed by these gigantic Aeolian harps. The strange thing is that each man hears his own melody and harmony, just as the wind blows.[32]

Strindberg and after him the entire *fin de siècle* doggedly pursue "this natural art, where the artist works in the same capricious way as nature, without specific aim."[33] The artistic expression of this capricious creation, a production that preserves the decrepit character of possibilities gone awry, is found in and through sound:

> I sought a melody for a one-acter [sic] of mine called *Simoom*, which is set in Arabia. To that end, I tuned my guitar at random, loosening the screws haphazardly, until I found a chord that struck me as something extremely bizarre, though without overstepping the bounds of beauty . . . Here is my melody, as it was composed by chance: G, C#, G#, B, E.[34]

Without overestimating the coincidental correspondence, but also not dismissing it, we should point out only that a few years later Schoenberg too will base his music on accidental series and combination of pitches whose sequence will make up a sort of genetic code of acoustics. It is sound in contortion that painfully wins for itself momentary consciousness. The impulse behind Schoenberg's sound-compositions remains the same throughout: to break open the stranglehold of symmetrical sound that enervates the senses, channeling them into preloaded and ideologically charged outlets.

This is how a new form of temporality is created—a type of duration that implants the distorted, uncomfortable but intimate psychology of belonging into sound.

FLORAL TRANSPOSITIONS

By now we are in the familiar place of misaligned perspectives that nonetheless serve as points of orientation—the supporting, primary elements of the *fin-de-siècle* world view. Would it surprise us that this topic is addressed by Arnold Schoenberg in his main theoretical work *The Theory of Harmony* (1911), in the opening paragraphs?

Schoenberg begins with the observation that, as much as 'our' time is in search of, what it has found most of all is comfort: "We understand today better than ever how to make life pleasant" and comfortable.[35] Comfort is a way of sweeping problems under the carpet. It dispenses with things that one cannot quite grasp. Seeking the reassurance of comfort, one inadvertently lives superficially, for "the prerequisite of comfort is: superficiality."[36] Such fundamental apathy makes a world view easily accessible, because one perceives only what is agreeable and without risk: "It is thus easy to have a 'Weltanschauung', a 'philosophy', if one contemplates only what is pleasant and gives no heed to the rest."[37] But it is the remainder, like the carnage of perceptions in Strindberg's railway carriage, one should be responding to. This is the only place of value, the type of value Schoenberg will find in the detritus of tonality that otherwise operates blindly, awarding to the consonant sound, or the pleasing, comfortable sonority, the uncontested leadership in acoustic construction. Hence, when all is said and done, one cannot exculpate oneself from "the rest—which is just what matters most."[38] Unlike measured taste, the dustbin of sensations is something that cannot be standardized and repackaged, as has been historically the case:

> In light of the 'rest' these philosophies may very well seem made to order for those who hold to them, whereas, in that light, the tenets which constitute these philosophies are seen to spring above all from the attempt at self-vindication. For, curiously enough, people of our time who formulate new laws of morality (or, even more to their liking, overthrow old ones) *cannot live with guilt*! Yet comfort does not consider self-discipline; and so guilt is either repudiated or transformed into virtue. Herein, for one who sees through it all, the recognition of guilt expresses itself as guilt.[39]

One is 'beyond good and evil' or beyond dissonance and consonance and encounters the world as a solution in waiting.

Closing off this brief overture that introduces the main argument *Theory of Harmony* develops, Schoenberg cites some examples of such an honest, creative practice, among others, bringing up the name of Strindberg, who could never settle for the numbness of comfort. All this reaches the, by now, familiar tone of jagged, razor-sharp perception, disowning "Comfort as worldview! The least possible commotion, nothing shocking [because] those who so love comfort will never seek where there is not definitely something to find."[40]

Even if the simplicity of the text seems to make the message somewhat trite or even forced, there is no doubt that the type of information it is trying to convey falls in line with the general profile of the times—with the need to continually respond to the unseen, mysterious, grotesquely degenerate plane that underpins culture and life in general. Here, we have the mechanics of deranged sensations in operation, where 'truth' is unclean and unpleasant, its outlines found in decompositions of stability. Only a change of state, a metamorphic transformation, uncovers something genuine through the jarring quality of the unfamiliar contrast.

The first decades of 1900, of which we are the intellectual descendants, were still a nascent period when the plasticity of modern experience could be accessed only indirectly. The descriptive power of language had not as yet managed to catch up with life. Paradoxically, this condition made such an experience more real, given how the body had to be involved and unfolded into communicative circuits, material and psychological. And this is perhaps why, as we will see, the only grammar accorded to this type of reality is a profound and unmitigated crisis of the word—namelessness and loss of language (*Sprachlosigkeit*)—a loss that can be expressed only in the psychometrics of resonance and the kind of imagination that it inspires, because only sound, its physics and logic, possesses the kind of immediacy other media of expression lack.

The physical self, caught up in the semiotic textures of the *fin de siècle*, is also implicated in a new biology where vision must employ the ear that opens the mouth. This strange, almost mythological, hybrid materiality and perception came to a point of highest intensity during the brief moment of *Jugendstil* and its ornamental configurations.[41] In an aesthetically pleasing form (hence the immediate criticism of *Jugendstil* as disingenuous), identity is captured in the density of hybrid shapes that redistribute the human form among stems, leaves, roots, and flowers. The simplicity of the human body in its stark nakedness is brushed into graphic formulas that, in their flow, turn everything into the wavy echoes of plant psychology—all along delimiting the field of phenomena as a space for the transplantation of new sensuality.

The somewhat inauspicious beginnings of the movement have their origins in a rhizome-like topography. The *Jugendstil* ornament is a parasitic invention that takes over spaces in the manner of ivy overgrowing a brick wall:

> It is of utmost importance, that the beginning of *Jugendstil* is built up through the ornament . . . But what type of an ornament? In book

> design it first emerges in vignettes, head and border-edges, initials. Soon it expands, incorporates picture motifs into itself, pulls and presses the letters, transforming the type, grows over the walls, wallpaper, carpets, posters, windows, captures furniture, baseboards, shortly even the street pavement. Such ornament knows almost nothing of the abstract repetition of the same, which up until then, served as the accepted, never broken principle of all ornamentation. Not one time an edge or a frame, uses exactly the same elements as another one; some kind of variation always takes place.[42]

We find here repetition of transplanted events, cultivating new, existential registers in their relentless drive for self-expression—in other words, all the elements that also constitute the instrumentality of Freud's and Schoenberg's mature output. The frame itself, the edges of its constantly shifting borders and not the contents, is the principle of construction: "It is the drive to frame everything that is more important . . . Picture and frame . . . the actual, framed paintings—begin, in fact, to mix one with the other."[43] And even if the contents still possess some autonomy, they do not simply fill out the picture, but rather are co-opted into the momentum of this 'enframing' construction, with the result that "Everything is mixed through, mixed together, water with plants, plants with swans or snake-bodies, and these with human appendages."[44]

The effect is more than purely aesthetic or thematic—the logical expression of artistic style and technique taken to the limit. The emerging point of contact between the force of abstraction and reality is much more serious—it presents itself as the between-zone or the in-between realm, where the human and the biological in all their plasticity meet surface to surface, exploding in a curious, a-signifying mythology, "this strange between-realm of plant-like, lower animals of the sea bottom."[45]

Jugendstil inadvertently represents nothing less than the structuring force of creation, its unprecedented 'organization' and genetic structure. It took until the middle of the last century to identify and reinscribe this dynamic force into the living tissues of the body, as the DNA code that connects us to the rest of organic life in the same flowing, ribbon-like manner. But before this could take place, before science could catch up with the new representational vocabulary that at one point simply becomes a latent part of our existential lexicon, there was this unquenchable need to search for 'man' outside of the human already launched by Nietzsche in such a spectacular fashion. Such investigative logic also underlines Schoenberg's fundamental necessity to search for sound outside of the musical sonority, or Freud's compulsion to search for psychology outside of the 'I', and then inadvertently return to it, proving that it does not belong to itself, and is not found where it thinks it is. The mechanics of this outbound search cannot help but to weaken and blur the dimensions of the body, whether it is the human body or the body of received traditions. As we will remember, the ego as projection of the body's surface is also only the echo of its constitutive dissolution. This is why *Jugendstil* receives and works with such "poverty of corporeality—more a notion, the feeling of a presentiment, than an exact reproduction."[46] But its sense of responsibility for this almost invisible, acoustic materiality means that *Jugendstil* overburdens the ornament with meaning that, in the end, ornamentation cannot support. There is no better evidence of this than its short-lived cultural presence and the type of opposition that it was soon to receive, like the one from Adolf Loos, for instance, who famously equated ornamentation and crime. This kind of negative reaction is not merely polemical. Rather, it exposes the fundamental flaw of conviction, found in the graphic element—something Loos will also oppose in architecture. This is why most of the *fin-de-siècle* avant-garde establishment attempted to distance itself from the too quickly and easily achieved comfort of vision and its crystalline, polished identity.

Nonetheless, *Jugendstil* does express the elementary 'organicism' of the *fin-de-siècle* modernity, radicalizing the network of its connections, responses, and investments, giving them the reality of a mundane life that is deceptively ornamental:

> All the shapes merge into each other, all becomes a mass of feeling, driven by the soul and moved 'organically'. The total, in this way *lived* ornamental world of *Jugendstil* is a type of materialization of the "hazy mist of the soul", as Maeterlinck says. It is exactly here where the uniqueness of *Jugendstil* lies, which separates it from all the earlier epochs and meanings of ornamentation. The abstract figures themselves, the simplest graphic elements, according to the *Jugendstil* theory, are always subordinated to feelings or 'experiences', in fact *inhabited* . . . [This is why], it is not the ornament that 'expressed' the soul-life of this strange time, but men themselves led the life of the ornament, the souls themselves became ornaments.[47]

Furthermore, such aesthetics, in a counterintuitive motion, discards the outside for the inner dimension, or at least it folds space, so that the two become practically indistinguishable. Nature is drawn into the ornament, literally and existentially. All this emerges in full force through the complete architectural design of living quarters condensed into the deceptive simplicity of a home (*Heim*). It is a total space, where individuality and architecture coincide, creating a strange fusion between the inner and the outer dimensions.[48] Of course, the attempted incorporation of the strange and the foreign into the space of domesticity is a contradiction in terms. When it is nonetheless carried out, the only result that one can be certain of is the uncanny feeling of the *unheimlich* that, as we already know, is defined by Freud as a strange relationship of distance and unfamiliarity one feels precisely in the most intimate and known surroundings.

It seems to be the case then that, notwithstanding its ability to recompose and recalibrate the elementary forces that otherwise would remain in the state of unconscious slumber, *Jugendstil* creates another shell in which it simply dresses itself. Eventually, it will have to again cover what it meant to expose in a crust of all too cozy comfort, thus silencing the voice it wanted to liberate:

> For indeed, the masquerade of the historical style was thrown-off, but only to be exchanged for the clothing of the new style, which now seemed to be tailored to the living body of the Zeitgeist, like second skin . . . It was the idea of *Jugendstil* to surround men, in fact the entire epoch, *in noisy reflections of their own insides*, to wrap them up in these moulds. Narcissus died because he had lost himself in his own image.[49]

The problem, as we already know from Hermann Bahr, was precisely vision that proved to be powerless to reformulate the organic textures of experience. Vision was not enough, and this is why the *fin-de-siècle* individual must learn how to listen.

Jugendstil ultimately retreated into the numbness and silence of the image. It became too comfortable with itself and the world, moving totally into the realm of comfort, feeling at home as it tried to take full control over the structure of life. It is perhaps not surprising that it should eventually buckle and fall into expressionism that stretched its horizontality vertically, thus once again accentuating the three-dimensionality of experience. Such vertical reasoning became the most important characteristic of Schoenberg's method of composition.

***

The crucial aspect of this aesthetics of renewal through dissolution is asymmetry. Asymmetry constitutes the most direct response to the unevenness of cultural experience and becomes the personification of the times, its mask that, as its etymology already implies, reveals as much as it hides in the processes of covering (persona=mask). The instability that goes hand in hand with asymmetry releases creative self-preservation.

This is how functionality and aesthetics, the predominant paradigm of the *Jahrhundertwende*, meet along the entire depth and breadth of the body—its knotted plasticity. This is also how the body must come out of itself and acquire shapes whose outlines are not immediately accessible. In other words, what must happen is a process of transition, or, even more radically, metamorphosis to another state—a state that can at least fain a certain level of stability in a reality whose first-order expression is numb consumption of impression.

And thus, in the final assessment, *Jugendstil* fails because its anaesthetized surface is allowed to take over almost completely:

> This is why, the so enchanted man, with a transformed face, becomes in his being, an organic, vegetative *soul*. The immediate body of bones, flesh, muscles, skin, nails and hair came to the *point of disappearance*, was dissolved in this driven seeing, and swallowed in this overall growth—"The body was nowhere to be found" . . . If the body, whose meaning as the soul's image, is not always a plant, it is surely each time, already something that transforms itself into one, because, turned around, the plants also are felt as animated [*be-seelt*], not only that, but also felt immediately as having a soul [*seelenhaft*].[50]

Transformation into mute organic structures is not simply soulless but a soundless reality, like Freud's death drive. And sound cannot emerge from silent germination: "An attempt to recognize and define the musical idea stands in clear contradiction to the sentimental poeticizing notion that a composition might arise from the motive as germ of the whole, as a plant grows from a seed. This is a childish notion."[51] This is how Schoenberg rebels against the standard musical 'organicism' associated with symphonic music, based on a continually cultivated grain of sound. Such organic arrangements create static forms, surrounded by the thick framing of tradition. It is teleological sound, acoustics of final resolution, imposed by the history of sound known as tonality.[52] And if there is one merit in *Jugendstil*, it is the fact that it at least started as an enactment of estrangement. It attempted to make contents unfamiliar precisely by disarticulating their spatial dimensions that could dispense with monolithic identities.

The best example of this procedure in sound that Schoenberg was soon to give up for the much more effective acoustics of liberated sonic material that is not attached to the stalk of rooted evolution is the song cycle opus 15, *The Book of the Hanging Gardens* (1908–1909), his first atonal composition based on a series of poems by Stefan George. The poems and Schoenberg's musical interpretation of their imagery are a typical example of literary *Jugendstil*, working in the spirit of defamiliarized nature, full of acoustics that almost parasitically overgrow the strictures and the standards of organic sound and word.[53] Just a couple of years later, in *Herzgewächse* (1911) and *Pierrot Lunaire* (1912) Schoenberg advances beyond mere estrangement and gives a real body effect to his acoustic shapes: "Both works transcend the Jugendstil out of which they arise. For the musical expression assumes an immediacy that no longer requires any metaphor: the music consumes the images and speaks directly from within."[54] In the end, *Jugendstil* pushes culture into the quietude of organic processes. Man and with him the voice of ideas disappear into the thick foliage of unawareness.

If the picture painted in such a radical way might seem too exaggerated, it is because modern scholarship tends to approach that time period through the overly strict objectivity of art in its historically specific style, thus divesting the phenomenon it intends to investigate of its living quality and the force of the overall project. Style is not an empty research indicator but a sociocultural phenomenon whose footprints are to be examined in line with its specific purpose and utility. At the turn of the previous century, art is practically synonymous with urgency and agency. Surely, it was also a stylistic manipulation, but its intricacies attempted to cross the dividing line between representations of life and its everyday reality.

Fin-de-siècle aesthetics became a mass medium of communication. As such, aesthetics was a lot more than a simple artistic tool of decadent expression, dissatisfaction, creativity, and so on. Rather, it launched a new semiotic code based on dissipation and disappearance that established the outlines of the formula according to which our modernity finds its point of orientation in its disembodied, technological reality. This is why Dolf Sternberger, already in 1934, saw *Jugendstil* as eventually imprisoning the senses. "Tragically the enchantment closed their [men's] eyes. What seemed like a solution to them was incarceration and condemnation."[55]

The final accusation leveled against *Jugendstil* by Sternberger from his vantage point of already some removal concerns what is really at stake in both the turn-of-the-century discourse and our theoretical elaboration of its silhouette—the crisis of sound, or rather its physical and polemical necessity as well as the anxiety about its potential lack. By his estimation, *Jugendstil*, which in the final analysis multiplies itself in a checkerboard pattern of silences, launches its own surface to a level where "muteness transforms itself into the mask of a cry."[56] *Jugendstil* failed to fulfill and attain the principle that the turn of the century needed desperately—that of sonority that, in its call, allowed one to pay attention and, more radically, to see what otherwise was inaccessible.

6 Psychology of Distances

The longing for form is "always taciturn and it always disguises itself behind many different masks", wrote György Lukács in 1910. He added, "Perhaps it would be a paradox to say that the mask is its form. But the mask also represents the great, two-fold struggle of life, the struggle to be recognized and the struggle to remain disguised."[1] This struggle is the symptom of a reality put to the test in the perceptual laboratory circa 1900.

Masking and 'cover-up' infiltrated architectural theory already in the middle of the nineteenth century. Most interesting were certain architectural principles, like *Bekleidung* or 'dressing', introduced by Gottfried Semper.[2] The technique of 'dressing' targeted aesthetic form inscribed in the raw materiality of buildings. It searched for art in the simple, engineered skeleton of an architectural structure. In this way, not unlike the ego's functional extension as the body, Semper created a method of enhancing surfaces that could *amplify* the repressed material qualities. But a crude application of ornamental mascara is not what he had in mind:

> Masking does not help, however, when *behind* the mask the thing is false or the mask is no good. In order that the material, the indispensable (in the usual sense of the expression) be completely denied in the artistic creation, its complete mastery is the imperative precondition. Only by complete technical perfection, by judicious and proper treatment of the material according to its properties, and by taking these properties into consideration while creating form can the material be forgotten, can the artistic creation be raised to a high work of art.[3]

The correct application of the mask, not unlike its use in classical Greek theater, should elevate the already extant identity, like the actor's character, making its psychological essence come to the surface with more intensity. This is how the material is forgotten through the perfection of form, whether it is the actor's own physicality, or the materiality of a structure. There is an intriguing connection here to psychoanalysis, given how it is also underpinned by the same logic of forgetting through the process of erosion or 'wearing away'.

Semper's influence on his peers is well documented. It was felt most acutely in the form of a productive hesitation, suspending architecture between pure engineering and art—a crisis that was perhaps best exemplified in the name architecture created for itself, like a new piece of clothing, becoming the 'art of building' (*Baukunst*). Already in 1849, Heinrich Leibnitz understood the relationship between building and art in these terms:

> We understand *architecture* to be that building activity capable of impressing the task of spiritual and moral (ethical) meaning on works arising out of naked need and mechanically assembled, therefore elevating a materially necessary form to an *art-form* . . . It is true that form is determined by mechanical theory and structural laws, similar to how the parts of a machine perfectly fulfil the function for which they are created, but form will forever remain rigidly constrained as long as it lacks the exterior habit, the characteristic that can elevate this *mechanism* to a living *organism* that can speak and signify its inner essence. This moment will occur only when the visual *art-form* places a transparent mask over this mechanical *core-form* . . . This characterizing element will be the creation of form or the *ornament* of architecture. Its purpose resides not in the building's structural functioning but, on the contrary, will articulate only symbolically the function of the core-form, precisely displaying all of its relations, and in this way endowing the work with that independent life and that ethical sanction through which it alone can be elevated to the work of art.[4]

The language of independent life through elevation and enhancement, where the mute, mechanical core is augmented and extended beyond its 'natural' parameters into organic, lived configurations, prefigures the elementary, symbolic dictionary of the *fin de siècle*. Schoenberg used it most effectively in "*Hertzgewächse*" (Foliage of the Heart)—a song composed in 1912 for *Der Blaue Reiter* almanac, edited by Wassily Kandinsky and Franz Marc. Based on the symbolist poem by Maurice Maeterlinck, the song not only attempts to understand the general human condition through sound (something Schoenberg will never give up) but also does so through interaction of surface and depth. The acoustic texture of a surface is precisely only a transparent mask through which the deep, musical core is made manifest. Only in this way can the spiritual architecture of modern life be stabilized and more importantly analyzed through expression.[5]

This is also why the body as such was used persistently by the turn-of-the-century 'expression-industry', including architecture:

> Architects sought assistance from the analogy of the human body: just as the body follows the form of skeletal construction, so architecture should reveal its inner truth . . . The metaphor of skeleton and flesh

appears in numerous writings on truth in architecture of the time—from Fritz Höber to Hendrik Petrus Berlage.[6]

A similar sentiment is expressed by Joseph August Lux, the author of the first extended monograph on Otto Wagner, arguably the most prominent and representative figure of Viennese modernism in applied arts. Writing in 1910, Lux concluded that "it is simply untrue that the bare skeleton is the final word in beauty . . . artistic form must be rediscovered in *the new elements*."[7] The search for these new elements was characteristic not only of the desire for organic architecture but also of the character profile of the entire cultural context. It was also brilliantly displayed by Otto Wager, whose projects and designs were impelled forward by wave after wave of modified materiality.

In *Moderne Architektur* (1902)[8] Wagner identified the two main architectural trends of the time: the symbolic one, originating with Semper, and the purely constructivist, initiated by the encroachment of engineers on the traditional space of building. These two trends were supported by the philosophy of realism and idealism, and professionalized into the identity-types described by Wagner as those "hermaphrodites of art and vampires of practice."[9] In opposition to these 'two species' of designers and theoreticians, Wagner constructed a type of *responsive materialism* as the most immediate expression of modern times, enclosed in the communicative texture of the modern city (*Großstadt*) and its relentlessly extended physical and perceptive limits. He came closest to the actualization of this philosophy in designing the railway system (*Wiener Stadtbahn*) for the city of Vienna, as well as the network of water regulation canals that meander through the city like a vascular system, reanimating its dimensions and thus giving life to the space inside their circumference. Responsive materiality was also found in the condensed monumentality of individual structures, which, as Fritz Neumeyer points out, can be understood as the actualization of a "body at work." This is how "the architectural object sought to achieve its cultural and urban function of assimilating, integrating, mediating, and finally *transforming* the given context of modern reality."[10] Thus, monumental appearance brought merely engineered spaces to life by accentuating the physical forces behind the artifice, bringing them to the surface, like the imprint of veins through the skin. These are the same forces that, at any one time, threaten to rip a structure asunder. This is how Wagner implanted life into construction—by suspending its exclusive professional activity, and instead remaking it into an aesthetic surface that mediates between the individual and the world at large.

***

In 1896 Georg Simmel, one of the most underappreciated figures in the history of social thought, presented the still enduring image of city life as

> An inner barrier . . . between people, a barrier, however, that is indispensable for the modern form of life. For the jostling crowdedness and the motley disorder of metropolitan communication would simply be unbearable without such psychological distance. Since contemporary urban culture . . . forces us to be physically close to an enormous number of people, sensitive and nervous modern people would sink completely into despair if the objectification of social relationships did not bring with it an inner boundary and reserve.[11]

Aesthetics is the system that regulates this boundary of distances and proximities. Clearly indebted to Riegl, Simmel understood cultural space as fluctuating perpetually between symmetry and asymmetry. Interchangeably and in varying proportions, symmetry and asymmetry serve as the ground for the expressive power of a civilization. Following this line of reasoning, any society oversaturated with the rationality that symmetry brings with it must eventually flip polarities:

> The origin of all aesthetic themes is found in symmetry. Before man can bring an idea, meaning, harmony into things, he must first form them symmetrically. The various parts of the whole must be balanced against one another, and arranged evenly around a center. In this fashion *man's form-giving power*, in contrast to the contingent and confused character of mere nature, becomes most quickly, visibly, and immediately clear. Thus, the first aesthetic step leads beyond a mere acceptance of the meaninglessness of things to a will to transform them symmetrically. As aesthetic values are refined and deepened, however, man returns to the irregular and asymmetrical. It is in symmetrical formations that rationalism first emerges. So long as life is still instinctive, affective and irrational, *aesthetic redemption* from it takes on such a rationalistic form. Once intelligence, reckoning, balance have penetrated it, the aesthetic need once again changes into its opposite, seeking the irrational and its external form, the asymmetrical.[12]

The pairing of individual and society is the latest, modern expression of this long-standing, dynamic asymmetry:

> Every epoch of human history seems to derive its unlimited number of manifestations from this dualism between movements of thought and life, in which the basic streams of humanity find their most simple expression . . . [The] present has found for this dualism the formulae of social versus individual, which draws its line through mankind and even through the individual man.[13]

Modern culture, according to Simmel, has come to the point of overlapping with *life* as such, more than ever before. This type of unprecedented organic pragmatism provokes a rebellion and even a denial of form as the principle of engagement with reality. Thus, an entirely new type of cultural experience is triggered—"a new phase of the old struggle [which is] no longer a struggle of a contemporary form filled with life against an old, lifeless one but a struggle of life against form as such, against the principle of form."[14] This was the same revolt against form that also found expression in abstract painting, blurred the dimensions of the human figure, psychoanalytically deconstructed identity, and released atonality as the blueprint for all sound.

This rebellion against clear definitions was the prosthetic technology the *fin de siècle* invented for itself—doing so most effectively through sound. The idiosyncratic quality of the aural element (the reason it was being harnessed to the entire expanse of communication networks) rests in its ability to accommodate contradiction on a mass scale. This is also the primary difference between the purified, iconic eye and the ear. Simmel says this unequivocally: "Transmission via the ear . . . actually incorporates contradiction. It forces a form that is intrinsically and sensorily directed towards unlimited number of participants to serve a content that completely excludes all of them."[15] This is how distances were suddenly inscribed into the modern sensory apparatus and its psychology, as opposed to being externalized, using the transcendental networks of metaphysical reassurance, or even something like the painterly perspective. Instead, these new, modern distances became aesthetic *values* in their own right.[16]

Thus, aesthetics found a new location in the *intervallic* quality of being, especially in the anonymity that distance offers. The point is, however, that aesthetics has always been and still was, according to Simmel, the basic procedure of how phenomena arise, how they organize themselves, and how we behave in relation to their cultural complexes. And the question that immediately suggests itself, the one that the *fin-de-siècle* generation asked in a series of modulations, concerns the authenticity of experience. This is how art entered the cultural scene full force, as the practice of perceptive manipulations that recalibrate distances between various phenomena. These manipulations of perspective give the modern individual an unprecedented flexibility of choice in communication that connects or dissociates us from the world at large. In the end, it seems that we must accept and learn how to inhabit the tension that cannot be calmed—the partially formed space lying between distance and proximity.[17]

This model captures the rudimentary process of modern life, especially its exaggerated demands of responsibility placed on individual identity. Closeness and removal, enhancement or delimitation of space—these are the elementary combinations of all communicative relationships. These work according to the paradoxical condition of estrangement that nonetheless connects and welds various elements into new configurations. Simmel eventually transfers this type of modeling to architecture and the figure of what I have described elsewhere as an 'inhabited ruin'—a place of reconciliation where repositioned identity and its continually disheveled organs of perception encounter each other.[18] This is how distance and proximity meld together in various proportions of sensitivity:

> For a very sensitive beholder, the peculiar distance between works of art and the immediacy of experience becomes especially clear when the object is very close. For less delicate perception, a greater distance from the object itself is required for the enjoyment of this charm of distance, as for example in stylized Italian landscapes or paintings of historical dramas. The less cultivated (and childlike) aesthetic feelings are, the more fantastic, the farther removed from reality the object must be by which the artistic work achieves its impact. A more sensitive viewer does not require such a materialistic prop. The artistic form of the object itself provides him with secret charm of distance from things, liberates him from their dull pressure, carries him from the realm of nature to that of spirit. He will experience this even more intensely when art deals with proximate, low and relatively secular material.[19]

This type of communication at a distance is also lodged in the body, in the raw logic of its material setup. Hence, it is not at all surprising that in another short essay, "On Aesthetic Quantities" (1903), Simmel proposed the following:

> The human figure presents an aesthetic miracle insofar as it maintains its aesthetic value through all possible enlargements and diminutions of scale. The reason for this is that its aesthetic proportions, with which we are in *solidarity*, take on such importance and concreteness for us, and have such immediate inner necessity, that they dominate everything else. Indeed, the human figure is perceived as a norm for the qualities and proportions of everything else: man is the measure of all things also in visual matters. When we are dealing with relationships among human beings, however, the problem of quantities arises again.[20]

This is why the body's shape is extended culturally, beyond its merely physical reality. This type of extension or more accurately amplification, because it is always, in some form, a speaking body, generates the psychological and the aesthetic textures that overlay the organic element, only to reveal and open it to investigation. Combined, these dimensions make up the physiognomy of hybrid formations, so ubiquitous in the first few decades of the last century. We have already seen this quite clearly in Freud's ego-psychology. The formation of identity's expressive power depends on the sonorous distances of words and body parts—distances that break through resistance, erode certainty, condensing and dissolving various forms of intimacy. Experience is given voice in one overarching project of making distances speak. All of this, especially when Simmel noticed the 'aesthetic resistance' of objects, just like Freud did the psychic resistance of patients, revolved around the nonexchangeable, core element of plasticity.

***

The project of subjective dissipation eventually took on a new measure. It was Walter Benjamin who saw it in the fullness of its unfolding, where the individual, blinded by exaggerated proximity to his own identity, lost the *tension* and the power of distance.

Benjamin's famous essay "The Work of Art in the Age of Mechanical Reproduction"[21] indicted modern culture for the loss of aura, or the enchanting distance that previously had been the domain of art:

> That which withers in the age of mechanical reproduction is the aura of the work of art. *This is a symptomatic process whose significance points beyond the realm of art.* One might generalize by saying: the technique of reproduction detaches the reproduced object from the domain of tradition. By making many reproductions it substitutes a plurality of copies for a unique existence.[22]

Such a field of sprawling reproductions degrades the authenticity of experience. By coming too close, we fake intimacy, elevating technological vision as immediately consumable satisfaction to the position of dominance.

Benjamin's argument is nuanced, and very often this essay has been read too quickly by commentators who see a straightforward relationship between technology and its negative impact, going off like an emergency light indicating everything that went wrong with modernity. The subtlety of what Benjamin proposed rests on finding a new scale for how perception must be adjusted to the modern, mass society. The latter must be dealt with not simply on the political or demographic plane, but also and especially psychologically: "The adjustment of reality to the masses and of the masses to reality is a process of unlimited scope, as much for thinking as for perception."[23] This is why the problem is also one of aesthetics, because only such aesthetic orientation provides the means to mass-produce a feeling of individuality with the cohesion of a crowd. Without such a scale, perception degenerates into a ritualistic exercise of administered sensuality. It is precisely this condition that Benjamin attempted to capture under the concept of the decay of aura:

> We define the aura . . . as the unique phenomenon of a distance, *however close it may be.* If, while resting on a summer afternoon, you follow with your eyes a mountain range on the horizon or a branch which casts its shadow over you, you experience the aura of those mountains, of that branch. This image makes it easy to comprehend the social bases of the contemporary decay of the aura. It rests on two circumstances, both of which are related to the increasing significance of the masses in contemporary life. Namely, the desire of contemporary

> masses to bring things "closer" spatially and humanly, which is just as ardent as their bent toward overcoming the uniqueness of every reality by accepting its reproduction. Every day the urge grows stronger to get hold of an object at very close range by way of its likeness, its reproduction. Unmistakably, reproduction as offered by picture magazines and newsreels differs from the image seen by the unarmed eye. Uniqueness and permanence are as closely linked in the latter as are transitoriness and reproducibility in the former. To pry an object from its shell, to destroy its aura, is the mark of a perception whose "sense of the universal equality of things" has increased to such a degree that it extracts it even from a unique object by means of reproduction. Thus is manifested in the field of perception what in the theoretical sphere is noticeable in the increasing importance of statistics.[24]

In a similar gesture to Simmel, Benjamin attempted to isolate a perceptive orientation with the power to shape distances, in the processes enabling a psychological stance that worked through the tension of contrasts. Tension and contrast were precisely the lost dimension when modern technology begun to artificially enhance experience exclusively as a form of proximity. But it is the ability to distance oneself from things that might be too close and bring closer the phenomena that practically lie beyond the range of immediate perception that matters. This acuity is not a matter of magnification or reduction, at least not in the way that technology achieves the feat, but a redefinition of spatial dimensions in their intervallic quality. By the time Benjamin wrote his essay in 1936 that power had been almost completely usurped by the mechanical eye of the camera lens and other techniques of projection. The example of statistics is exemplary here. Statistics create an interesting paradox, given their uncanny ability to magnify the individual as the function of a population. They engineer a collective measure of simultaneous presence, constructing an artificial social sense that functions almost on par with the eye and the ear. A statistic magnifies by giving the other senses a mistaken outlet, and it is this kind of magnification that creates the false security of closeness. Benjamin's archetypal example of this kind of logic is film, where the body of the actor is taken beyond the edge of dissipation:

> With a vague sense of discomfort he feels inexplicable emptiness: *his body loses its corporeality, it evaporates*, it is deprived of reality, life, voice, and the noises caused by his moving about, in order to be changed into a mute image, flickering an instant on the screen, then vanishing into silence.[25]

Even if we might want to disagree with Benjamin about the artistry involved in film production, the point is that cinematic consumption reconfigures the experience of perception enough to make a regressive step in the application of attention through physical presence. Instead of *being* a stressed body, what the *fin-de-siècle*'s version of vanishing and disappearance still preserved, *having* an appendage-like body in its overexposed, graphic reality increasingly became the norm. Interestingly, taking a purely visual approach, Benjamin accused psychoanalysis of a similar transgression because "the camera introduces us to unconscious optics as does psychoanalysis to unconscious impulses."[26] Optics is not at all inconspicuous in relation to the ego as we know, but it is not the driving force behind psychoanalytic perception, which belongs to the aural realm. Nonetheless, the issue remains that of attention:

> Fifty years ago, a slip of the tongue passed more or less unnoticed. Only exceptionally may such a slip have revealed dimensions of depth in a conversation which had seemed to be taking its course on the surface. Since the *Psychopathology of Everyday Life* things have changed. This book isolated and made analyzable things which had heretofore floated along unnoticed in the broad stream of perception. For the entire spectrum of optical, *and now also acoustical perception*, the film has brought about a similar deepening of apperception.[27]

Writing in the 1930s, Benjamin's hindsight look at the ebb and flow of history of the immediately preceding decades and the end of the swell initiated by the *fin de siècle* is already skewed by the oncoming tidal wave of a new, ideologically sealed world. In only three decades political ideologies matured and 'learned' how to operate on the basis of politics as aesthetics, harnessing the logic of renewal and remaking of all relationships, taking it out of the laboratory for 'reality testing', and transplanting it into the laboratory for universal destruction.

Already in 1909, the Italian futurist Filippo Tomasso Marinetti called for the glorification of war, "world's only hygiene, adding,

> We will destroy the museums, libraries, academies of every kind . . . we will sing of the multicoloured, polyphonic tides of revolution in the modern capitals . . . Art, in fact, can be nothing but violence, cruelty, and injustice.[28]

What in the first decade of the twentieth century was an isolated call of poetic derangement became a commonly available imagery by the time Benjamin wrote the conclusion to his essay in 1936. The message was similar, but this time spurred by real events, such as Italy's Ethiopian campaign, that also set Marinetti's pen into motion.

INHABITED RUINS

The power to forge new distances needed its own psychology of ruination.

Or at least this is Simmel's proposal. The 1911 essay on the ruin (*Die Ruine*) is an experiment in perception that begins in a familiar place: "Architecture is the only art in which the great struggle between the will of the spirit and the necessity of nature issues into real space."[29] Ruins combine the antagonistic forces of nature and intellect, embracing the precarious stability of hesitation. They extract the essence of that tension that expresses something of the already lived duration, as well as its final expiation that is still to come.

The process of ruination produces hybrid topographies where unification of eroded and worn-out elements and constant variation in their composition is put into practice. This is how identity lives the inexhaustible contradiction of cohesion and dissipation, experiencing a 'mysterious harmony' not at all that different from Schoenberg's deconstruction of sound, based on the technique of perpetual variation. In fact, we might want to consider how the image of a ruin would apply to Schoenberg's acoustic universe. Are Schoenberg's compositions not a form of decay, on all levels, overfilled with sound that ruins itself as it finds an expressive outlet?

The intimacy of ruins is deposited in the mechanism of incessant transition:

> The ruin of a building . . . means that where the work of art is dying, other forces and forms, those of nature, have grown; and that out of what of art still lives in the ruin and what of nature already lives in it, there has emerged a new whole, a characteristic unity.[30]

Ruins are hybrid spaces, sites of partially collapsed physical and perceptive dimensions. This is where space is experienced in the full vibrancy of its communication, where mutual infiltration of culture and nature takes place. But it is not an act of willful aggression "because destruction here is not something senselessly coming from the outside but rather the realization of a tendency inherent in the deepest layer of existence of the destroyed."[31] In this sense, the ruin is "nothing merely negative"[32] but the most intense, material expression of life's plastic energy. This is how the ruin "creates the present form of a past life, not according to the contents or remnants of that life, but according to its past as such."[33] The past transposed in such a forceful manner into the present can result only in an aesthetic form "where there is no longer any sharp division between perception and thought."[34] In other words, ruins are not simply dilapidated, dead structures. Just like words are different lengths or frequencies of the otherwise limitless reality, ruins in which we dwell, unlike structures in a mute state of disrepair, are echoes of the paradoxical completeness of life. This is the case because "nature and culture are only two different ways of looking at one and the same thing."[35] That thing is the process of wearing away and the liberty of asymmetry.

How can we not see Freud in all this? His psychoanalytic practice triggered by the wearing away of memory, initiating the process of psychological collapse of enclosures supported by complexes claiming exclusivity over personal identity? This is how the work according to the past as such begins. It is a treated past, whose plastic force has been isolated and analytically applied, not the past attached to the dust jacket of commemoration.

Psychoanalysis is still the best-known psychological laboratory for reshaping cultural identity. Entering the strictly choreographed premises of Freud's practice at Berggasse, number 19, the patients were very much mimicking ruins of the past in the present that now stood in the analytical configuration like postcards of decaying monuments. They also entered a debris field. Freud's consulting room was laden with ancient artifacts. More than in a manner of speaking, in Freud's space, one was forced to encounter a living ruin—a blueprint for what must be done with the psyche, digging it up from the recesses of time immemorial and exposing it to 'natural' psychological forces. In this way, the ruin became a sort of incantation of time in its actual construction—a point of contact and transference from the otherwise untouchable, distant realities. The power of expression given to what is otherwise mute lies, like the ruin, somewhere between the 'not-yet' and the 'no-longer.'[36]

***

Whereas in Simmel's view, man and culture can still search for reconciliation in the pragmatics and dynamics of the ruin as an aesthetic principle, everything seems already ruined for Benjamin, gone too far into the amnesia of an unresponsive mass: "Even the most perfect reproduction of a work of art is lacking in one element: its presence in time and space, its unique existence at the place where it happens to be."[37] Temporality and placement are what inhabited ruins still have in abundance.

The *fin de siècle* was steadfastly designing such a placement, and with it a new sense of duration. Otto Wagner's modernism, for instance, intuitively reintroduced subjectivity into the actual construction of spaces as points in active tension. Ultimately this kind of style created architecture whose frozen choreography engineered buildings as mimes of cultural life.[38] In other words, Wagner's *Baukunst* attempted to solve the problem of mass as merely dead weight. Buildings must be engaged with, like one would engage a landscape or, indeed, a person face-to-face—inescapably. This solution is similar to Simmel's resolve to alleviate the tension splitting the individual and society. The only remedy is aesthetics:

> What drives the modern individual so strongly toward style is the unburdening and veiling of the personal, which is the essence of the style. *Subjectivity and individuality have reached the point of collapse*, and in the stylized making of form—from social manners to the furnishing of a dwelling—resides an appeasement, a toning down of this acute personality to a general idea and its law. It is as if the ego could

> no longer support itself solely, or at least no longer wished to reveal itself, and so it puts on a more general, more typical, in short, a stylized garment . . . Stylized expression, the way of life, taste—all are limits of ways of distancing, in which the exaggerated subjectivity of the period finds a counterbalance and a mask.[39]

The mask, in fact, is a regulatory system of distancing. Style might, in some measure, mistreat perception, but its masking logic is absolutely indispensable for the creation of an aesthetic psychology in a communicable form. Performed correctly, such an aesthetic psychology expresses a new architectonics of the senses that escape the neurosis of overmagnification.

Benjamin's carefully constructed case against such overmagnification and its reproductive technologies rests on our lost ability to calibrate distances, to match them to the most significant events, phenomena, and problems facing a culture at a specific time. Instead, contemporary society is vulnerable to all the distortions put into practice by the seemingly transparent technologies of perception. Like Simmel, Benjamin considers architecture the archetype of such a tension between perception and distortion, contemplation and distraction:

> As regards architecture, habit determines to a large extent even optical reception . . . For the tasks which face the human apparatus of perception at the turning points of history cannot be solved by optical means, that is, by contemplation, alone. They are mastered gradually by habit, under the guidance of tactile appropriation.[40]

Architecture is the original medium of mass communication. This is where we find encoded perceptive attention, in various compositions—the same attention that as the twentieth century wore on has been eroded completely to the benefit of distraction. In this respect architecture and its habit-forming topography are similar to film. Art becomes a matter of collective appeasement and comfort in escapism: "Art will tackle the most difficult and most important [tasks] where it is able to mobilize the masses. Today it does so in film."[41]

The relationship between attention and distraction is the problem whose exact distribution the *fin de siècle* took upon itself to solve, even if that role was simply presented to it by default, as a historical necessity. It attempted to do so through a curious blend of forces that cultivated and implanted a new habit-forming space of collective perception. This is what architecture proper tends towards as a matter of course and what aesthetics, across the many communicative fields, is compelled to invent by finding ways to inhabit the contradiction of stability found in controlled dissipation.

In short, the search concentrated on collective aesthetic identity—on locating intimacy in distance that could shift the registers of meaning. Only such *mobile* space promised to create enough momentum for interpretation and analysis. We have already found it in Bahr's *Augenmusik*, in *Jugendstil*'s organic morphologies, in the body that let itself be absorbed by the stress of its own physical form. It was a body that, the way it was presented by Schiele, exposed the inherent voraciousness of all forms of coherence. All these ingredients, when fused together, activated the tension and the creative energy of incongruence and misalignment.

Otto Wagner's architecture is again exemplary on this account. For instance, his design for the *Ferdinandbrücke* created an overall effect that invoked "an image that is meant to explain by visual analogy the inherent [material] forces and tensions. The pylons endow the construction with power; they pretend to be the 'actors and performers' whereas the steel beams appear to have a passive role of reacting to forces."[42] Wagner did not simply construct buildings but shaped spaces—expressive locations as theaters of engagement where distance and intimacy are allowed to take up residence.

Hermann Bahr championed Wagner's aesthetics for precisely that reason. Contrasting Wagner's personality and philosophy of space with the traditional Viennese style, Bahr brought the character profile the *fin-de-siècle* project to the forefront, especially its attempt to derive a rejuvenated aesthetic communication. The symbolism and the phenomenon of the mask played an important role in Bahr's contrasting description of Wagner versus the city. On the occasion of the architect's seventieth birthday, Bahr wrote,

> The tragedy of the Viennese is constituted in this, namely: they are never allowed to have a natural face. The genuine man in Vienna has been forbidden in the last two hundred years . . . And our eternal Viennese problem is whether, in the end, we will finally have enough strength and courage to take-off the masks and show to ourselves who we are! And no one, among those who presently live in Austria, endured and persevered in this task in a more marvelous way than Otto Wagner. Otto Wagner is the opposite of Vienna's Ringstrasse. Whereas everything on it speaks to effect, Wagner is all about expression; where there is arbitrariness, here one finds necessity; there swindle, kitsch, theatre, here always merely what the matter requires.[43]

Eleven years earlier, in 1900, Bahr already wrote about the experience of walking along the Ringstrasse:

> If you walk across the Ring, you have the impression of being in the midst of a real carnival. Everything masked, everything disguised . . . Life has become too serious for that sort of thing. We want to look life in the face. This is what we mean when we talk of 'realist architecture', that is, that the building must not only serve its intended purpose, but must also express, not conceal, that purpose.[44]

To do this one needed a different range of sensitivity, the type theorized, for instance, under the label of *Einfühlung*—the feeling of mutual resonance between an individual and an object—originally translated into English, somewhat misleadingly now, as empathy. The operative principle here was perception *at a distance*, most famously proposed as a type of aesthetic orientation by Theodor Lipps and Wilhelm Worringer.

In *Abstraction and Empathy* (1908), Worringer extended Lipps's theory by arguing that aesthetic perception should not be exclusively based on feelings of pleasure or displeasure. *Einfühlung*, the feeling of unity between phenomena and imagination (this is Worringer's addition), underpinned the psychology of everyday culture, as it carried the entire network of historical styles forward. It was a history of how past cultures had connected with their world through abstraction, invented to assuage our primary dread of space. By necessity, at the beginning of the twentieth century, one searched for reassurance in active transitions, even into the inorganic state:

> All transcendental art sets out with the aim of de-organicising the organic, i.e. of translating the mutable and conditional into values of unconditional necessity. But such a necessity man is able to feel only in the great world beyond the living, in the world of the inorganic. This led him to rigid lines, to inert crystalline form. He translated everything living into the language of these imperishable and unconditional values. For these abstract forms, liberated from all finiteness, are the only ones, and the highest, in which man can find rest from the confusion of the world picture.[45]

Worringer's historical topography of artistic expression, based on Riegl's work, was suspended between moments of geometric abstraction (like the Egyptian pyramids) and empathetic immersion in reality that blended seamlessly with the environment. The point is that "this dualism of aesthetic experience . . . these two poles are only gradations of a common need, which is revealed to us as the deepest and ultimate essence of all aesthetic experience: this is the need for self-alienation.[46]

Like most of his generation, Worringer grappled with the problem of a space that could create its own realm of experience. In the end, the type of theory he proposed was implicated in the incalculable effects of intimacy at the threshold of dispersion: "Our knowledge of phenomena is complete only when it has reached that point at which everything which seemed to be a boundary becomes a transition, and we suddenly become aware of the relativity of the whole."[47]

What we, then, find under the name of art at the turn of the century is not simply a creative/artistic product of 'pictures at an exhibition' but a model of universal relations without a proper name. Art was chosen because it could adjust to and direct the constantly changing circumstances, currents, tastes, and fashions. In fact, we could even say that the entire *fin de siècle* was a *flimsy* reality that anticipated the impulsive, technological acceleration that was to define the twentieth century. This is how modern aesthetics smashed the standards of all proportion, becoming a *point of contact* with the formlessness of culture. Art, as this unprecedented blend of forces, simply became the first technology of active interface with the increasingly disembodied presence. This is again evident in Wagner's polemical arguments about *Baukunst*:

> There are thousands and thousands of things that modern culture has devised, and for many of them art today has already found forms—many have even been given a perfect form. They do not recall the forms of past times; they are completely new because their premises and their essential principle have issued from our very own ambition and perception.[48]

Thus, besides setting out on a form-finding mission, art became a corrective device, ontologically speaking, with the architect as the most progressive modern social type: "The architect with his happy combination of idealism and realism has been praised as the crowning glory of modern man. Unfortunately he alone feels the truth of these words, while his contemporaries stand off to the side, little interested. I too, at the risk of being accused of a delusion of grandeur, must join in the song of praise."[49]

All this revolved around *practice*, which, similarly to psychoanalysis, produced functionality as new alignments of space and identity. One telling sign of how far-reaching such a reconstructive project must be was the inclusion of practically every artistic and engineering expertise in the comprehensive architectural 'tool box', including acoustics:

> The architect is able to project any space in such a way that its acoustics can be predetermined with apodictic certainty. The theory guiding this determination is quite simple and runs as follows: make the length of the sound waves as equal as possible and avoid all reflections.[50]

The desire for total control and predictability of even something as diffuse and unwieldy as sound speaks to the demand placed on experience. These were also the elementary functional conditions that both psychoanalysis and atonal sound found indispensable, as they tried to impress the design of their own psychic and musical spaces onto the shape of cultural identity.

MYSTERIOUS HARMONY

"A surface is not simply a geometric composition of lines [but] a certain distribution of the sensible."[51] Redistribution of the sensible formed the core of the *fin-de-siècle* culture, consistently straying into the limit of what, on the surface, should be allowed.

How do we, for instance, reconcile Adolf Loos's poor hearing and his long-standing friendship with Arnold Schoenberg? What is it that attracted a practically deaf architect to music, especially Schoenberg's music? H. H. Stuckenschmidt in his biography of Schoenberg recounts this incident noted by the composer:

> Loos comes to the rehearsal. Is very nice. Especially to Webern. He finds a similarity between himself and me in that I make alterations in my compositions at rehearsals. But this is wrong: I don't work out the composition during the building, as he says, but on the contrary such alterations are very rare with me and are mostly only concerned with notation marks. But I didn't contradict him, because he was very amiable; and then, above all, he is so deaf that one can hardly talk to him; it would be exhausting for me to contradict him.[52]

But there was a deeper connection. It seems that the nature of this almost grotesque attraction of a partially deaf architect to such an intricate auditory experience was the principle of construction itself—the same one on which Loos's practice was based, combining elements that practically should have never belonged together. Schoenberg redesigned sound along the same lines of asymmetry, incongruence, and inversion. He thus liberated sound's raw plasticity, using it without aesthetic prejudice to remodel sonic materials into unprecedented aural phenomena. And this fell in line with Loos's notion of value, because "There are no precious and cheap materials in the world . . . To God and the artist, all materials are equally valuable."[53] Schoenberg expressed the same opinion about sonic value enclosed in the structure of the chord. There are no precious or discredited, consonant or dissonant sounds, only lived harmonies that carry the identity of a musical structure forward, especially when it is freed from the constraining standards of tonality. And the erasure of tonality created musical forms without proper names, practically identical to the type of cultural anonymity espoused by Loos. Its adage was inconspicuousness—the modern individual, according to Loos, should stand out as little as possible in the fray of culture.

Sound, liberated from the strictures of tonality was, in a way, silenced. Hence, another intriguing correspondence to the austerity and muteness of Loos's buildings suggests itself. The façade or the skin of the building must be freed from all superfluous information, such as ornamentation, so that life on the inside can resound that much more persistently. Famously proclaiming that ornamentation in the modern world is a form of degeneracy, Loos's modeling of culture transposes life, as much as possible, to the inside of a dwelling—cultivating the intricate textures of intimacy. Outward display of identity constitutes a perceptive disorder. And Loos pushes this paradigm as far as it will go, searching for and finding a similar type of experience at the architectural heart of music making–the concert hall.

The solution to the problem of 'good sound' is unusual, if not uncanny. Acoustics is condensed as memory in the physical identity of building materials. In other words, it has no placement in topographical configurations "because the acoustics of a concert hall is not a question of spatial solutions but a question of materials."[54] Raw, organic materials such as wood, not unlike a sound recording, absorb and store the physical imprints of previous encounters and events. This is how music performed in a concert hall leaves a resonant, molecular memory in its walls:

> Have our ears changed? No, but the material used to construct the hall did change. The material, which, over the course of forty years absorbed constantly good music, has been impregnated with the sounds of our orchestra musicians and the voices of our singers. These are the mysterious molecular transformations that until now we could only observe in the wood of a violin.[55]

Such responsive materialism became the aesthetic language of modern culture, because "Materiality must be again deified [and] matter is downright mysterious."[56] The philosophical lining of pan-materialistic ethos is prominent in the historical moment of Loos's architecture, Schoenberg's music, and Freud's psychoanalysis. All three, whether it is through the ego's ultimate dwelling in the body, molecular restructuring of sound, or architecture as a type of organic engineering, elevate matter to the position of an ontological principle: Everywhere lives only materiality. Moreover, "material, substance, product and fabric need a certain time of molecular rest"[57] that can be communicated only through the tension of frontiers and the intimacy of distances.

For instance, the muteness of Loos's façades—the smooth surfaces of his buildings, which caused scandals and accusations of the gross prostitution of culture (as indeed was the case with the famous Michaelerpltaz house in Vienna), walls devoid of any superfluous elements—deliberately attempted to distance perception from the cumbersome weight of traditional communication. The narrow, innocuous doorways, slim windows with restricted access, difficult to use as viewing points, are only some of the many instances of this type of thinking through design. Its best example is Loos's last built project—the Müller Villa in Prague. The villa achieves perfection in topographical redirection, especially in terms of sound—the outside is completely silenced, for the benefit of intimacy and resonance on the inside.

Such design philosophy reproduced the most elementary of psychological conditions—reversal of perception, or, as Freud described it, the complete turning around of the subject's own self (*Die Wendung gegen die eigene Person*):

> Such turning of an instinct [drive] upon the subject's own self is made plausible by the reflection that masochism is actually sadism turned round upon the subject's own ego, and the exhibitionism includes looking at his own body . . . The essence of the process is thus the change of the *object*, while the aim remains unchanged. (SE 14: 127)

Loos's *reversed* design unmoors the subject from standardized topographies, so that he or she can feel the presence of subjectivity that much more powerfully on the inside. Structurally, this is expressed as so many psychological reconstructions of identity through space—a principle also fundamental to how Schoenberg approaches the entire edifice of composition.

Therefore, a house, designed from the inside out, is a *cultural assemblage*, the "concrete means of communicating the experience of space"[58]—the blueprint model of relationships in general. It is instinctual architecture that occupies this difficult-to-define border region between psychology and the physicality:

> an instinct appears to us as a concept on the *frontier* between the mental and the somatic, as the psychical representative of the stimuli originating from within the organism and reaching the mind, as a measure of the demand made upon the mind for work in consequence of its connection with the body. (SE 14: 121–122, my emphasis)

Only this kind of *frontier mentality* could find a way of redefining the threshold of experience.[59]

The orientation in which "the house does not have to tell anything to the exterior"[60] violates the traditional identity of spaces as either exposed or hidden. Loos worked on escaping both of these extremes, proposing instead sequences of transition zones. His structures wedged themselves into the surroundings, whether urban or natural. The house as living quarters was the pinnacle of such transitivity—a space of a continually retested presence that enables a new type of an emotional economy, efficient in its communicative interfaces. In this form, architecture matched modern psychology: "The silence that he [Loos] prescribed is not more than the recognition of our schizophrenia: the inside has nothing to tell the outside because our intimate being has split from our social being. We are divided between what we think and what we do." And crucially, such a division can be bridged only through listening.

There is a difference between hearing and listening. Nietzsche, for instance, thought that the scientific spirit, immortalized in the figure of Socrates, heard but refused to listen. Socrates incessantly silences his interlocutors by hearing without listening (this is his stratagem), barring any possibility of a dialogue, despite the enactment of its mechanics. To practice listening, one must know how to instill periodic silence. This is the point at which the psychoanalytic session culminates, when analysis is broken off, or begun again from a different point of contact:

> If "to hear" is to understand the sense (either in the so-called figurative sense, or in the so-called proper sense: to hear a siren, a bird, or a drum is already each time to understand at least the rough outline of a situation, a context if not a text), to listen is to be straining toward a possible meaning, and consequently one that is not immediately accessible.[61]

Loos's architectural paradigm, in the fullness of its program, is based on this type of listening, leading the architect to conclude that "the image of a genuine structure when pictured on paper makes no impression. I am extremely proud when the interiors I've created have completely lost their effect when photographed; when the inhabitants of my spaces do not recognize their own flats in photographs."[62] The graphic posturing of photography that cannot represent Loos's spaces at all (his interiors are remarkably 'unphotogenic') creates something uncanny or *unheimlich*. Significantly enough, Freud proposed the uncanny as an *aesthetic concept*, based on disquieted perception, where home (*Heim*) is suddenly experienced through estrangement. The effect is amplified awareness and, in the end, a reinforcement of intimacy, even though this type of estrangement is triggered by the feeling of alienation from things most familiar, "for this uncanny [*unheimlich*] is in reality nothing new or alien, but something which is familiar and old-established in the mind and which has become alienated from it only through the process of repression" (SE 17: 241). Such a feeling is a psychological experiment of sorts, a test of one's physical surroundings devised by the ego: "The whole thing is purely an affair of *reality-testing*, a question of the *material reality* of the phenomena" (SE 17: 248, my emphasis). Any architecture worthy of the name, whether in Loos's or Wagner's hands (the differences between them notwithstanding), is first of all the practice of reality testing. Once again, all this speaks to the analytical habitat of the *fin de siècle*, its staged *mise-en-scène* where "the classical distinction between inside and outside, private and public, object and subject becomes convoluted."[63]

***

Most studies exploring cultural history take music to be simply an epiphenomenon of the prevalent artistic currents of the time, even if acknowledging the special challenge of grasping such musical reality. As Walter Frisch observes,

> Even in such writers as Carl Schorske or Roger Shattuck, music tends to remain undernourished—sometimes abused—perhaps because its basic language and "meaning" are notoriously difficult to decipher, let alone relate convincingly to some broader zeitgeist.[64]

A much higher-yield approach is to consider sound as carrier of the encoded logic of experience—a sort of sonic depository of its elementary principles that, unlike other dimensions, makes enough noise to be heard through the acoustic symptom itself. This approach is valid, if only because there is a preponderance of evidence that music, or at least sound theory and its metaphorical value, was used as signifying matter by the practitioners of modernism, both well-known and forgotten, like August Endell, who in turn influenced the much better-known Kandinsky.[65] This 'wire-mesh' of sound to which empirical realities in their full complexity were fastened had been prepared in the nineteenth century. This is when acoustics experienced its fully mature scientific phase, along the way enabling a new type of investigative methodology based on sound—consistently extended to the entire spectrum of physiological, psychological, and sensory dimensions. We may only mention here Hermann von Helmholtz[66] or even the theory of consonance and dissonance proposed by Theodor Lipps.[67] In other words, acoustics and the plasticity of sound were used as a new approach to an old problem: how to perceive things that would otherwise remain inaccessible. This is how sound assumed the technical role as the amplifier for the unconscious, the signifier behind the theory of color and architectural spaces, and the principle of communication orchestrating the exchange between the body and its psychology. In other words, resonance became the plateau for a new point of contact with reality. The materiality of sound, its phonology, served as the lining for the identity-forming discourse. Historically, it was the framing of sound into musical form that was invested with cultural capital. That dominance was broken around 1900, when sound's power to express dissipation and formlessness of centered identities was unleashed.[68]

The Italian futurists and the Noisism/Brutism movement constitute the exclamation points of this cultural history of sound. The idea was to weld the senses together by appending them to each other and forming hybrid if not mangled complexities. Carlo Carrà, for instance, in "The Painting of Sounds, Noises, and Smells" (1913), called for the 'sonorification' of the visual field: "We Futurist painters maintain that sounds, noises and smells are incorporated in the expression of lines, volumes and colors just as lines, volumes and colors are incorporated in the architecture of a musical work."[69] In the same year, Luigi Russolo, in his "Art of Noises", castigated Western art music and called for the infiltration of the standard musical tone by raw, environmental sonority: "Now we are satiated and WE FIND FAR MORE ENJOYMENT IN THE COMBINATION OF THE NOISES OF TRAMS, BACKFIRING MOTORS, CARRIAGES AND BAWLING CROWDS THAN IN REHEARING, for example, THE "EROICA' OR THE "PASTORAL."[70] The "art of noises" also made its way into poetry, as is evident in Dada—a nonsensical word, which, as Tristan Tzara in his 1918 "Dada Manifesto" provocatively confirmed, "MEANS NOTHING."[71] Among the many examples of "sound poetry" that made use of noise and gibberish "freeing the word to explore its 'innermost alchemy'"[72] was Kurt Schwitters's *Ursonate*.

The Dada movement originated in Zurich at the end of World War I:

> At the first Dada evening, on July 14, Huelsenbeck read a poem with 'real sounds' . . . The poem becomes a static object that can be read from all sides at once and, in fact, in several languages at once, like 'The Admiral is looking for a house to rent', read simultaneously in German, French, and English.[73]

Now, if we recall Anna O. and her almost instantaneous rendering of texts into foreign languages, could we not say that she was really the first Dada poet? This could, in fact, be the reason why in 1928 André Breton and his fellow surrealist Louis Aragon celebrated fifty years of the invention of hysteria: "We surrealists insist on celebrating the fiftieth anniversary of hysteria, the greatest poetic discovery of the latter nineteenth century." They added, quoting Joseph Babinski—a French-Polish neurologist whom Breton admired, and who, together with Freud, was a star student at the famous Charcot clinic at Salpêtrière "When an emotion is sincere and profound, and it stirs the human soul, there is no room for hysteria. And in that we have the best so far that we have been given to learn."[74] Hysteria, or rather its creative energy, became the poetry of the real.

At its most condensed point, such poetry is but an echo of the thin-skinned psychology of the modern individual. It encapsulates the generative impulse of perception, especially its ability to break apart signifying structures all the way down to their raw elements, such as phonemes or tones. Thus, August Endell was able to write in the following manner about the unknown art whose practice should focus as much on the immediacy of perception as on the underlying living fossils of encoded impulses:

> At present, we still have little in the way of form-art; that is, form-images [*formale Gebilde*] that are nothing and mean nothing in themselves, that affect us directly, without any intellectual mediation, like *the tones in music*. This type of effect [*diese Wirkungsweise*], this new art is almost unknown, and the little of it that does exist is dismissed with a scowl.[75]

Such form-art, like a genetic code, redesigns the boundaries of what counts as communicative experience. And the sequential release of these 'drifting dislocations' is how the modern subject increasingly began to encounter the social environment. The psychic calculus of continually *dislocated forms* underlies Freud's psychoanalytic theory, as well as increasingly underpinning the artistic, philosophical, and scientific complexes.

Theodor Lipps referred to such formative logic as the 'rhythmic-psychic happening'. In this he anticipated Freud's characterization of his brand of psychology, activated by the mechanism known as *Wiederholungszwang*, or the compulsion to repeat:

> Rhythm possesses an all-encompassing meaning inside the psychic life. In general, rhythm is a specific type of succession of particular psychic processes of the same type . . . I can also say that it attempts in the same manner to rhythmicize [*zu rhythmisieren suchen*] the entire psyche. As such, I am considering rhythm, I have to repeat this again, in an all-encompassing sense, namely as the overall characteristic way of the psychic occurrence or its discharge.[76]

Hence, it is only in the contorting power of vibration that the modern psyche recognizes itself. Placed among the noise of mutually sounding elements, Lipps describes it as 'psychic resonance':

> Empathy [*Einfühlung*] is at the same time the empathy in sounds and sound-connections as well as the empathy in colours. Every sound, every sound-connection, every colour, or more precisely the psychic process of perception of these, has its own "rhythmicity", which becomes or can become the rhythm of the overall personality.[77]

Lipps's argument anticipated the *frontier* psychology of primal instances and last resorts that produced a new kind of space as form, a space "treated by a technique that may be defined as matter and as movement."[78] The topography of this space described the *eviction route* taken by standardized cultural enclosures and meanings. More accurately, signification in its abstract and physical registers was deflected away from its usual indicatory values, thus releasing the intensity of evocation, streaming it into new types of cultural presence. In the end, it was sound that best fulfilled the function of a materiality that had its own range of experience and the uniqueness of its own time line—a time line that accentuated the deficit of abstraction.

***

Schoenberg's music, in its successive reformulations, whether it is the decayed romanticism of the *Gurrelieder*, the atonal expressionism of *Erwartung*, or the tattered rationality of the twelve-tone compositions, encompassed the entire expressive horizon of the *fin de siècle*. All these dimensions were present together in the cultural psychology of that time, as they also dominated individual identities, Schoenberg's included. Thus, in a diary begun on January 19, 1912, and kept for the next twelve months, we find this entry:

> Perhaps it will be better if I compose again now. Or: it is better. I remember writing a poem ten to twelve years ago in which I wanted to be old, unambitious and quiet. Now when I suddenly see the old possibilities of unrest again, I almost have a longing for them. Or are they still there? . . . But in the next few days I will certainly compose . . . I

> know where this comes from: "Spring", always my best time. I already feel things stirring in me. *In this I am almost like a plant.* It is the same every year. I have nearly always composed something in the spring.[79]

This plant-like psychology, the elastic frontier between the self and the environment, can take many forms and shapes, according to the complexes at work in each, and the particular combination of expressive dimensions. It can outgrow, or overgrow itself, hide in the thicket, camouflage its profile, only to display that much more aggressive individuality later, like van Gogh's sunflowers. The point that needs to be stressed again is the different range of materiality that reordered the *fin-de-siècle* perception. It was a materiality that reinvented the conditions for psychological and cultural presence—its basic unit of analysis and construction was the inversion of physical and abstract limits. This is still the contradiction at the heart of modernity—the compulsion to always begin anew, from a higher plateau of decomposition, if not a complete failure. Today, it constitutes the ethos of the everyday struggle, invested in the overcoming of the self on the road to personal success. A hundred years ago this tension of never concluded transitions was being invented through the interminable Freudian analysis, or Schoenberg's acoustic form of inhabited ruin in sound.

Transitions are the communicative interfaces between stability and destruction. Schoenberg's sound is full of them, as was his personal life—moments that, in the end, become inseparable. In a letter to Alma Mahler-Werfel, he wrote,

> If I am to be honest and say something about my works (which I do not willingly, since I actually write them in order to conceal myself thoroughly behind them, so that I will not be seen), it could be only this: it is not meant symbolically, but only envisioned and felt. Not thought at all. It meant something to my emotions as I write it down . . . I don't want to be understood; I want to express myself—but hope that I will be misunderstood. It would be terrible to me if I were transparent to people.[80]

This psychological paradigm of opaque presence is a form of self-distancing. Only it could provide a different range of awareness and some practical stability through incomprehensibility.

In 1926, Universal Edition (Schoenberg's main publisher) printed a short manuscript entitled *25 Years of New Music*. Schoenberg contributed an introductory article to the publication. The topic for reflection was tonality:

> It is in fact a question of economy. According to my feeling for form (and I am immodest enough to allow this alone to have control of my composition) the playing of even one tonal triad will bring its own consequences and demand a certain space, which cannot be allowed inside my form. A tonal chord *arouses expectations* of what is to follow, and,

> working backwards, also has influence on everything that has gone before it; one can't demand that I should overthrow everything that has gone before it because a triad has by chance been introduced.[81]

We are back on familiar territory here. Sound configured through tonality overregulates distance. Tonality binds by association, instead of opening up paths through which sound could find its own trajectory—a type of an association without commitment that the Norwegian psychologist Kristian Aars described right at the turn, in 1900:

> It is certainly convenient for certain psychologists to derive expectation from associations. But associations are accomplished facts [*vollzogene Tatsachen*]. Everything is [already] contained in an association. There remains nothing to expect.[82]

Instead, expectation according to Aars rests on 'association-lines' that enable experience without representative centrality. It is an expectation that blurs the certainty of vision to the advantage of the *aural element*:

> Expressed abstractly, we can say that expectation rests on an association between the present and the future condition of being conscious. But such association, in the time of expectation, is not factual. As soon as it is a fact, expectation is no longer there . . . Described through its own property, expectation is a potential for association [*Associationsgepräge*] . . . [Hence] expectation certainly does not rest on association, which always comes as a succeeding stage, or to give an illustrative example, it rests on the existence of *association-lines without an image*; on the same characteristics of experience, which lead towards its completion.[83]

Dissonance, the way it was treated by Schoenberg, was precisely this always renewed potential for association. The consonance-dominated tonality, on the other hand, filled musical spaces with slag heaps of accomplished events that buried the uncertainty of expectation under so many forms of predictable resolution. This drive towards dispersion of centers and identities was evident in Schoenberg's much discussed, innovative, and unusual orchestrations—an idiosyncratic combination of instruments that made sound not simply musically but also topographically. In fact, the surface of sound that Schoenberg so meticulously designed was a space one needed to enter and not simply experience aurally. This was the only way of encountering the harmony of no resolution, where the blending of demarcations among various dimensions outmaneuvered the power of regulatory systems.

7 The Anatomy of Namelessness

At the turn of the twentieth century, the old struggle of abstract principles for a system of coordinates that could bestow a new meaning onto cultural presence came to a point of exhaustion. The exclusive duels between idealism, romanticism, or even historical materialism, bouts of absolute spirit with the world, their still powerful presence notwithstanding, were displaced by an even more radical challenge against the fabric of meaning as such and the principle of form itself, as we have heard Simmel observe. And this in turn meant a persistent confrontation with the material structures of life—its many anonymous elements. In other words, the great systems of thought and knowledge, building and adding the plateaus of existence together into the recesses of dialectical and metaphysical infinity, collapsed under their own weight. This is how the construction site of 'our' modernity in which we seek placement to this day merged with the hypermodern science, delving into the incessantly smaller particles of space in search of the hiding place for universal forms that must not and cannot be challenged. In the early twentieth century the ground was prepared for physics to eventually split the atom. Today biology is catching up, fragmenting the human figure into ribbons of genetic code.

Around 1900, the paradigm of fragmentation took hold in the most radical manner. It was a historical and cultural version of a child's play whose 'innocence' tears off the wings of a butterfly to satisfy its curiosity. This is why the psychology of childhood took on such an importance, especially in psychoanalysis. But perhaps the most enduring and visible effect of the fragmentation paradigm is the category of individuality through which everything else had to be apprehended. The artist became the archetype of this subjective net, for the first time presenting a widely accessible model for an identity that was responsible only to itself through subjective fulfillment. Moreover, by releasing the immanent contrast between culture and its form, the artist acted almost in the manner of a natural force—inescapably. It is precisely through art and its exasperating calls for yet another project of renewal by rescaling all relationships that the sequence of anatomical lessons in perception becomes identifiable as what might be called a history of cultural transvaluations. Of course, the lesson in anatomy as the access point to the blueprints of life was not a new procedure—it had been practiced in various forms for centuries. New was the application of its principles, in the most literal way, to live phenomena, right across the physical and cultural spectrum. One example of the origins of such an attitude was Émile Zola's experimental novel—a type of analysis of culture through writing, begun in the 1880s.

FRAGMENTED SELF

Zola proposed to adapt literature to the medical paradigm, practically piecemeal. As his blueprint he took Claude Bernard's *Introduction to the Practice of Experimental Medicine*, observing,

> I really only need to adapt, for the experimental method has been established with strength and marvellous clearness by Claude Bernard in his "Introduction à l'Étude de la Médecine Experimentale". This work, by a savant whose authority is unquestioned, will serve me as a solid foundation.[1]

Bernard is famous for making vivisection the centerpiece of his method:

> If we wish to attain the exact conditions of vital manifestations in men and animals, we must seek them not in the exterior cosmic environment, but rather in the organic internal environment . . . But how can the interior environment of complex organisms such as men and the higher animals be known, if not by descending to them, in a sense, and penetrating them through experimentations applied to living bodies? This means that in order to analyze the phenomena of life, it is necessary to penetrate living organism with the help of vivisectional procedures.[2]

One sliced through living tissues and events, carving reality to size. Just how literal this crisscrossing of paradigms was can be surmised from Zola's conclusion that "it will often be but necessary for me to replace the word 'doctor', by the word 'novelist.'"[3] The ultimate justification for transplanting the body ripped apart by investigation into the space of a literary encounter *as* the novel rested on the understanding that "medicine, in the eyes of a great number of people, is still an art, as is the novel."[4] And as we already know, these two moments of medicine and the novel came together in a particularly spectacular fashion in psychoanalysis, especially through the personality and practice of its author.

Zola painstakingly applies Bernard's method to the art of the novel, emphasizing the crucial aspects of this practice that, in its raw form,

> treats [various phenomena] of practical experiments on living beings, of vivisection, of the preparatory anatomical conditions, of the choice

> of animals, of the use of calculation in the study of phenomena, and lastly of the physiologist's laboratory.[5]

It is this type of *literature as laboratory*, in which vivisections of cultural life and character experiments can be conducted, that Zola hopes to establish, convincing others that this is the most valid paradigm of literary engagement. In this form,

> Science enters into the domain of us novelists, who are to-day the analyzers of man, in his individual and social relations. We are continuing, by our observations and experiments, the work of the physiologist, who has continued that of the physicist and the chemist.[6]

Shortly thereafter, psychoanalysis takes up the position of a science that made the analysis of man into the cornerstone of contemporary culture and history. As such, it heeded Zola's call to operate not merely on the skin and bones but also on the passions and the intellect of man:

> In one word, we should operate on the characters, the passions, on the human and social data, in the same way that the chemist and the physicist operate on inanimate beings, and as the physiologist operates on living beings.[7]

Freud himself, besides entitling one of his 1932, *New Introductory Lectures* "The Dissection of the Psychical Personality" (SE 22: 57–80), compared the work of the analyst to that of a surgeon. And Schoenberg applied a similar procedure to sound. Stripping the skin of tonality released the deep character of the acoustic makeup, in the process capturing the nameless psychology of human passions. Anonymous compositions without key signatures and unnamed operatic characters, like the woman protagonist in *Erwartung*, are the figures populating Schoenberg's acoustic universe. Blending of the senses, paradigms, and techniques is the rule here, although physiology takes precedence, because ultimately it is the human body that must give up its ghost, its mystery:

> If the territory of the experimental doctor is the body of man, as shown in the phenomena of his different organs both in their normal and pathological condition, our territory is equally the body of man, as shown by his sensory and cerebral phenomena, both in their normal and pathological condition.[8]

In this sense, the experimental novel, psychoanalysis, and Schoenberg's acoustic forms that produce new and above all *living* sonic organs based an unprecedented anatomy of the tones in their body-like structures[9] are the completion of human physiology on the cultural scale:

> the experimental novel is a consequence of the scientific evolution of the century; it continues and completes physiology, which itself leans for support on chemistry and medicine; it substitutes for the study of the *abstract* and the *metaphysical* man the study of the *natural* man, governed by physical and chemical laws, and modified by the influences of his surroundings; it is in one word the literature of our scientific age, as the classical and romantic literature corresponded to a scholastic and theological age.[10]

This type of knowledge, in the manner of Simmel's ruin, works at the intersection of matter and intellect, never completely departing from either.

Zola kept a sustained attack on metaphysical thinking and the idealist writers of his generation. He tried to recuperate and reestablish the logic of materiality as the standard by which all reality was to be approached—a trend that came to an especially high pitch in architecture, as we have seen. In other words, from the middle of the nineteenth century onwards, art and philosophy engaged in a sustained project for the *extraction of materiality* from the shell of symbolic mystification, which until then laid claim to the pinnacle of understanding. This is how identity destabilized its own form, expecting of itself more than cultural phenomena were in the habit of giving. Experimental novel, the psyche, or 'atonal' sound became probationary spaces, where phenomena tested their limits through continually reinvented material textures:

> Respect for material! Respect for craftsmanship. Schoenberg had had these feelings from his youth onwards. He loved to work with leather, canvas, paper and string . . . As a carpenter he made book-shelves, tables, chairs . . . He cut curtains from material, stretched them on a wooden frame and painted them. He tinkered and sketched . . . His interpretation of the concept of material separates him radically from Busoni. For Schoenberg material is the note which is still unformed, for Busoni it is the unusual sound.[11]

Busoni at one point proposed that we admire Mozart as an innovator and seeker, but not as the forger of great musical material, prompting Schoenberg to respond in contradiction "Yes! . . . The material which is destined for the shaping of timeless things, attracts it; it is worthy of honour for us. It is the material vestibule of the spirit."[12] In this, once again Schoenberg displays a very similar type of materialism actively prompted by Adolf Loos, who called for a new reverence of the material world, on its own terms, without the added layer of complication and significations known as ornamentation, which leads only to corruption and a stylistic enervation of creativity. Zola's organic materialism is also unequivocal:

> Being obliged to make a study of man, we feel we cannot deny the efficacy of the new physiological truths. And I will add that the novelists

> are certainly the workers who rely at once upon the greatest number of sciences, for they treat of them all and must know them all, as the novel has become a general inquiry on nature and on man. This is why we have been led to apply to our work the experimental method as soon as this method had become the most powerful tool of investigation. We sum up investigation, we throw ourselves anew into the conquest of the ideal, employing all forms of knowledge.[13]

Hence, this inescapable conclusion:

> The metaphysical man is dead; our whole territory is transformed by the advent of the physiological man . . . In short, everything is summed up in this great fact: the experimental method in letters, as in the sciences, is on the way to explain the natural phenomena, both individual and social, of which metaphysics, until now, has given only irrational and supernatural explanations.[14]

Robert Musil expressed similar vivisectionist sentiments. These start in 1898 in his diaries with the "Pages from *monsieur le vivisecteur's* Book of the Night." *Monsieur le vivisecteur* is Musil himself, overflowing with the ability to straddle experience in a ballet-like dexterity, fusing almost seamlessly with the surrounding spaces—the capacity that only night possesses:

> I live in the Polar region, for when I go to my window I see only still white surface that serves as a pedestal for the night. Around me, there is organic isolation, I rest as if under a 100m thick cover of ice. A blanket such as this, gives to the eye of one who is thus comfortably buried a certain perspective known only to him who had placed 100 meters of ice over his eye. This is how it looks from the inside looking out—and from the outside looking in? I remember a fly I once saw interned in rock crystal . . . By being sealed within an alien medium it lost the detail of what one might call its "fly personality" and appeared to me only as a dark surface with delicately shaped appendages . . . *a new sensation of self emerges* that is like stepping suddenly in front of a mirror that has not received a single ray of light for days and, drinking in greedily . . . Recently I invented a very fine name for myself: "monsieur le vivisecteur" . . ."Monsieur le vivisecteur"—that's who I am! My life: the wanderings and adventures of a vivisectionist of souls at the beginning of the twentieth century! What is m.l.v.? Perhaps he is typical of the human being to come.[15]

The type of experience Musil longs for and describes is based on the plasticity of the ego. This is the same elementary plasticity that Freud isolated, enabling him to operate on experience itself. Psychoanalysis found a way of inhabiting the ambiguity of transitions, *reforming* the relationship between the inner and the outer, distant and proximate spheres of engagement into folded topographies of the "not-external outside and the not-internal inside."[16]

***

The irresistibility of the unknown is, of course, not new. It was the motivating force behind the institution of religion, its mimicry of experience through ritualized mannerisms. Institutionalized religion organized life according to a symmetric table of events, behaviors, and transgressions that erased man's myth-making powers. This was the main motivation behind Nietzsche's rage against religious perspective that, according to him, also invented the arid, scientific mono-lingualism. In contrast to this restricted dimension, one must look at reality and its dominant scientific mode through the perspective of an artist, who does not simply live for art, but *is* a living, artistic creation in his own right,[17] for it is art that ultimately spins out the network of significations that envelop the world.

"The existence of the world is justified only as an aesthetic phenomenon", wrote Nietzsche already in 1872, anticipating and contributing to the shape of culture around 1900. This type of aesthetics, as we might already suspect, has precious little to do with contemplation and reverence of beauty. Rather, Nietzsche is concerned with the meaning of aesthetics as a force that provides the energy for the creation of life in new perceptive dimensions. This is why:

> the aesthetically sensitive man stands in the same relation to the reality of dreams as the philosopher does to the reality of existence; he is a close and willing observer, for these images afford him an interpretation of life . . . This joyous necessity of the dream experience has been embodied by the Greeks in their Apollo, the god of all plastic energies.[18]

A link to Freud here is impossible to miss:

> The beautiful illusion of the dream worlds, in the creation of which every man is truly an artist, is the prerequisite of all plastic art . . . In our dreams we delight in the immediate understanding of figures; all forms speak to us; there is nothing unimportant or superfluous.[19]

In fact, at one point in *The Interpretation of Dreams* (1900), Freud refers directly to Nietzsche: "We can guess how much to the point is Nietzsche's assertion, that in dreams 'some primeval relic of humanity is at work which we can now scarcely reach any longer by a direct path' (SE 5: 549). Moreover, dreams are the testing ground for the "trans-valuation of all values", as Freud says, quoting Nietzsche's famous expression to refine his understanding of the psyche (SE 4: 330).

"At bottom, the aesthetic phenomenon is simple . . . to see oneself transformed before one's own eyes and to begin to act as if one had actually entered into another body, another character."[20] It is the famous Dionysian moment in its ecstatic unmooring of self that provides the opportunity to experience such a transposition—a process that later becomes the refuge and the energy source of psychoanalysis. Only in this kind of transformation do we reach

> Our highest dignity in our significance as *works of art*—for it is only as an *aesthetic phenomenon* that existence and the world are eternally justified—while of course our consciousness of our own significance hardly differs from that which the soldiers painted on canvas have of the battle represented on it.[21]

And what now speaks through such a dispossessed self is the anonymity of the universal:

> Insofar as the subject is the artist . . . he has already been released from his individual will, and has become as it were, the medium through which the one truly existent subject celebrates his release in appearance.[22]

This is why Nietzsche ultimately concludes that our highest dignity is lodged in the self that loses the vestiges of individuality, especially its exclusivity, for the benefit of anonymity. The ego is the *echo* of the world, not its artificial center. Such a disinvested self, a body stripped of its cultural and idealistic crust, was the mainstay of Greek tragedy. Here, the otherwise highly placed men in the echelons of the polis shed their civic/social 'portfolios' and experienced life through the otherness of a different skin, playing roles and entering realities as far removed from their station as possible—becoming slaves, for instance. This is how the body became the channel for other forms of communication, enhancing its own physiology beyond its limits, being "at once subject and object, at once poet, actor, and spectator"[23]—precisely the kind of identity that the *fin de siècle* worked tirelessly to implant into the modern cultural psychology. This is how the fight to regain the mystery, complication, and unfathomable depth of nature as a partially inhabited realm was being conducted. Once again we encounter a Freudian moment in Nietzsche's thought, because "man, to a large extent must be stripped of culture so that he can glimpse the truth."[24]

This truth is not simply located in the rawness of the instinctual life but in action that reclaims the sensory realm in its authentic shape of *perceptual instability*. And all this takes place through the aesthetic experience that combines culture and materiality into new psychological complexes. Truth, such as there is, simply acquires the character of an aesthetic artifact, underscoring the importance of what Nietzsche calls the Dionysian moment in history (even if it is largely fictionalized). The latter is found in the power of perceptive liberation, its ability to invent "a new world of symbols [where] the entire symbolism of the body is called into play."[25] It is a world of activity and experience, but not knowledge, given how "knowledge kills action; action requires the veils of illusion."[26] This very astute observation comes during Nietzsche's discussion of *Hamlet*. Such analysis anticipates a very similar conclusion reached by Freud in *The Interpretation of Dreams*, another indication of influence on Freud's thinking: "Hamlet represents the type of man whose power of direct action is paralyzed by an excessive development of his intellect" (SE 4: 265).

The fact that Nietzsche eventually found this active existence without knowledge in tones is due to the unmatched elasticity of sound—these physical phenomena without the body, according to Schopenhauer,[27] that propel things through space as "the immediate idea of life, [whereas merely] plastic art has an altogether different aim: beauty and eternity of phenomena."[28] In fact, Nietzsche, who during his life composed more than seventy works, such as songs, choral compositions, symphonies, and piano pieces, displayed an unreserved admiration for Chopin, and still practiced the instrument two hours daily after his breakdown, was as much a composer as, at any one time, a philosopher.[29] And what he found in sound is the opposite to the limpid plasticity of purely visual events. Their depleted elasticity cannot match the sonic phenomena's sense of a space without limit that literally brings the music to the eyes (Bahr's *Augenmusik*). One is roused to pay attention through this antidote of sonority. Given that we tend to be especially vulnerable when the availability and surety of knowledge are shaken, a condition that psychoanalysis works untiringly to promulgate as the indispensable step towards a 'cure', one is forced to look behind the otherwise blinding appearances as nonnegotiable categories. And whereas there is quite a bit of joy found in such an Apollonian stance as Nietzsche refers to it, because it is based on the illusion of optimism, "Dionysian art, too, wishes to convince us of the eternal joy of existence: only we are to seek this joy not in the phenomena, but behind them."[30] The imperative of looking behind inadvertently liberates dissonance as the logic of communication:

> It is precisely the tragic myth that has to convince us that even the ugly and disharmonic are part of an artistic game . . . But this primordial phenomenon of Dionysian art is difficult to grasp, and there is only one direct way to make it intelligible and grasp it immediately: through the wonderful significance of *musical dissonance* . . . The joy aroused by the tragic myth has the same origin as the joyous sensation of dissonance in music.[31]

Dissonance amplifies the indecisive and hesitating life but with enough density to be experienced as an identity of longing that extends the limits of possibility:

> We now understand what it means to wish to see tragedy and at the same time to long to *get beyond all seeing*: referring to the artistically employed dissonances, we should have to characterize the corresponding

> state by saying that we desire to hear and at the same time long to *get beyond all hearing.*[32]

Such a desire to get beyond hearing and seeing is the moment when we begin to listen with entire, organic surfaces, not only the isolated senses. This is precisely how Schoenberg proceeded in his project to redesign sound's physical dimensions, amplifying the sonority of what is left over, the sonority of ruins, after everything stylized and added on had been stripped away.

This acoustic resurfacing of reality renders environments and spaces discontinuous, loosening their geometry, so that the elements that make them up take on a different configuration. By resequencing events and juxtaposing them to each other (juxtaposition literally refers to being side-by-side without connectives), one is given the opportunity to *design* communicative pathways and not simply receive dispensed information as the philosophy of truth, the aesthetics of beauty, or the politics of cultural presence. This condition is exemplified in the profound crisis of the word that gripped the *fin de siècle.*

LANGUAGE WITHOUT A NAME

> In brief, this is my case: I have completely lost the ability to think or speak coherently about anything at all. First, I gradually lost the ability, when discussing relatively elevated or general topics, to utter words normally used by everyone with unhesitating fluency. I felt an inexplicable uneasiness in even pronouncing the words 'spirit', 'soul', or 'body' . . . Rather, the abstract words which the tongue must enlist as a matter of course in order to bring out an opinion disintegrated in my mouth like rotten mushrooms.[33]

According to Michael Huter, Hofmannsthal "was not only *the* poetic genius in Vienna at the end of the century, but also one of the most sensitive theorists of language, and among the first experts on Nietzsche to be taken seriously."[34] Such a sensitive, Nietzschian moment is most famously expressed in a short piece of prose from 1902, entitled "*Ein Brief*" (A Letter). As the title suggests, it is a fictional letter written by a certain Lord Chandos to his mentor, the philosopher Francis Bacon, who inquires about his protégé's literary silence. What follows is one of the greatest instances of doubt about language as the abode of meaning.

The *Chandos Letter* was described by Hofmannsthal as "one way into existence"[35]—an existence that provides a way out of the intellect and all these words. As such, it leads towards a space that, having lost its contemplative identity, merges with life itself. This is how we search for the out-of-place reality, locating it in the watering can, a cellar full of rats, or a spider's web, etc., as Chandos indeed does as he attempts to find an adequate description of the kind of condition that came one day and changed his entire perceptive disposition:

> Since then I have led an existence so empty of intellect [*geistlos*] or thought [*gedankenlos*] that I fear you can hardly form a conception of it . . . words fail me once again. For what makes its presence felt at such times, filling any mundane object around me with a swelling tide of higher life as if it were a vessel, in fact has no name and is no doubt hardly nameable . . . A watering can, a harrow left in the field, a dog in the sun, a shabby churchyard, a cripple, a farmhouse—any of these can become the vessel of my revelation.[36]

If Chandos now finds himself in the intellect and thought-free ('*geistlos*' and '*gedankenlos*') zone, a world that suddenly became expressive but practically uncommunicative, he is also falling back into the namelessness of the productive force behind all things and phenomena. The struggle for this new organic form of belonging is not easy, for it involves the entire physical apparatus of the body:

> I feel a rapturous [*entzückendes*], utterly infinite interplay in me and about me, and none of the interplaying essences is such that I could not flow into it. It is as if my body consisted entirely of coded messages that unlock everything for me.[37]

Such unlocking renounces the focus of perspective and evidence of the senses in favor of total, membrane-like resonance with the world:

> Everywhere I was in the midst, never confronted with mere appearance. Or I had the feeling that everything is metaphor, every creature a key to the others, and that I was the person who could take hold of them one at a time and unlock as many of the others as could be unlocked.[38]

All the usual categories of opposing tension, such as thought and body, culture and the animal world, loneliness and community, melt away into a state of a hybrid space and its hyperactive consciousness.

There is hardly a better example of the linguistic-emergency gripping the networks of communication circa 1900. Hofmannsthal confronts us with something like the 'anorexic corporality' of language. One becomes disgusted with words. Language acquires a transparency and flimsiness that threaten to destroy it. Its power to shape perception becomes inverted, and its cellular-semiotic texture implodes from within: "For people are tired of talk. They feel a deep disgust with words. For words have pushed themselves in front of things."[39] The skin of language, like a putrid body, smells of decay.

How can we reconcile this statement with psychoanalysis and the 'talking cure'? If we return to our previous discussion of Anna O., we realize that the psychoanalytic procedure, of which Anna is the starting point and prime example, is precisely to make the patient, if not disgusted, then defiantly exhausted by the constant stream of words. This state is a point of rapture, so that what hides behind words can come to the surface. The primary symptom of Anna's condition, as we will remember, is the loss of language.

Hofmannsthal is very much continuing (if not bringing to completion) Nietzsche's project of linguistic eviction, which is already clearly formulated in *The Birth of Tragedy*:

> Language can never adequately render the cosmic symbolism of music, because music stands in symbolic relation to the primordial contradiction and primordial pain in the heart of the primal unity, and therefore symbolizes a sphere which is beyond and prior to all phenomena. Rather, all phenomena, compared with it, are merely symbols: hence *language*, as the organ and symbol of phenomena, can never by any means disclose the innermost heart of music; language, in its attempt to imitate it, can only be in superficial contact with music.[40]

Hofmannsthal writes in almost exactly the same cadence, although the phenomenon of music is replaced by soundlessness and the plasticity of gesture (dance, mime, etc.). At first glance, this seems to constitute a complete reversal of Nietzsche's position, but that is not the case. Dissonant sound in Nietzsche and silence in Hofmannsthal are *metaphoric technologies* that construct a system of transfigurations, transporting the subject beyond the range of immediate sensory presence. Hence, Nietzsche's conclusion that "for a genuine poet, metaphor is not a rhetorical figure but a vicarious image that he actually beholds in place of a concept",[41] and in any case, concepts are simply 'stale metaphors', according to Nietzsche. In other words, metaphors are not merely literary devices but *literal* things that spin out of themselves the empirical world in divergent ratios to what had been prepared culturally. With this axiom Hofmannsthal is very much in agreement, both intellectually and creatively.

Nineteenth-century linguistic philosophy used the paradigm of metaphoric construction to study the reality of language. It proposed a close link between the body and the mind, where metaphors would not be simply poetic/figurative embellishments but literal phenomena. Exemplary here were such thinkers as Gustav Gerber (1820–1901), who proved to be of major influence on Nietzsche. Gerber not only published *Das Ich als Grundlage unserer Weltanschauung* (*The 'I' as the Basis of Our Worldview*, 1893), a title that is very interesting in light of our previous discussion of Freud, but also established the idea of language as an aesthetic construct, or an art form. This thesis is explored most directly in Gerber's 1871 book, *Die Sprache als Kunst* (*Language as Art*). Language for Gerber is a *Sprachkunst* (similar to the *Baukunst* in architecture) whose root is a sound-image or *Lautbild*, as the first building block of words, which is then extended metaphorically to more and more distant signs of the original *Sprachwurzel* (root). Taking this into account, Nietzsche is then able to say not only that metaphors are alive and that truth operates according to the aesthetic/metaphorical quality but also that all concepts are simply stale and overused metaphors that must be continually renewed.[42]

Hence, the fight against the usurping power of *stale* language responds to the pressing need to redress the increasingly overwhelming paralysis of culture instigated by the strange contraption of the individual as the expression of the social. The paradoxical nature of individuality as the principle of social organization is its always missing, expressive link to the presumed ownership of its own reality, and we can ask further, taking into consideration the beginning of our century, whether that problem has been solved. Assuming for a moment the idea and the phenomenon of individuality are inevitable, the problem that immediately arises is the construction of a space where individuality could thrive. This was precisely the task Adolf Loos took on as the problem of architectural design. Loos derived the blueprints of this space from the inverted topographies of presence, where space as limit could be continually challenged. Individuality, to support itself, needs the physics of unlimited potentials. These are found in space as total environment, where communication is equivalent to working with a chisel, continually sculpting the topography of cultural engagements.

Hofmannsthal, especially in his dramaturgical works, designed an experiment of mediation between the expectation for individuality and the inevitable presence of natural forces. Theater as ontological testing ground is not new to us. Perhaps the most extensive instance of its reach is Brecht's carefully designed mechanics of estrangement (*Verfremdungseffekt*). Hofmannsthal's theatrical practice, at this still early stage, was much less politically conscious, simply setting into motion the conditions for the two dimensions of the individual and society to collapse into each other.

If, as Benjamin Bennett points out in his comprehensive study of Hofmannsthal's oeuvre, "the idea of ordinary social life as an unacknowledged mysticism corresponds to the quality of the theatre as society",[43] then it is only a short step from such a conclusion (which inevitably suggests itself) towards the realization that what emerges here is a new form of community. "Form is a mask", says Hofmannsthal predictably, "but without form [there is] neither giving nor taking from soul to soul."[44] This new form of the social corresponds to the most interesting and evocative analyses of its space that emerge in the latter part of the twentieth century, such as Bourdieu's habitus or Goffman's dramaturgical blueprint for social interaction. But we can be even more radical here, and say again that all this effort at reformulation of relations in Hofmannsthal and in the *fin de siècle* not only concerns a new format of the social but also a society that attempts to become conscious of the physical, binding force holding together the structure of events. This is the reason why individuality became such a point of tension—because it stopped being simply a problem of integration into the most natural and genuine society (as it was in Rousseau, for instance) and became a new mass-scale phenomenon through which modern society distinguished itself from the previously defined coherence. *Zusammenhang* or coherence was of principal importance to Schoenberg's conceptualization and technical embodiment of sound. Before being replaced with the notion of the 'musical idea' in the 1930s, coherence was the driving force behind Schoenberg's understanding for how acoustic phenomena are to be generated through developing variation and the 'basic shape' (*Grundgestalt*) of sound. The latter eventually took the form of the famous row in Schoenberg's twelve-tone compositions.

Schoenberg has been often accused of a willfully precocious attitude that simply opposed standardized musical understanding for the sake of the pleasure found in rebellion. That this was not the case can be surmised already in 1917. When considering coherence and coordination, Schoenberg makes clear that they are not accomplished in a Dada-like manner where, as he amusingly writes, "the table salt screams in high-swung wrinkles" but through a different form of tonal individuality, based on repetition and the division of essences. It is a type of an *acoustic morphology* based on acoustic surveys of shared parts "neither too large nor too small for our sense perception."[45]

Of course, individuality, whether social or acoustic, is not exclusive to modernity. Socrates, as Nietzsche points out, is perhaps the first historical example of such a psychological configuration at odds with the aggregate, social life of the polis—a stance for which, not insignificantly, he is eventually put to death. What is new and what emerges at the turn of the century is the positioning and placement of individuality at the center of mass communication that paradoxically at its core cannot be otherwise than narcissistic and self-serving. And, if at the time of Socrates the individual was a privileged type who, under the special name of a philosopher, engaged in unencumbered pursuit of knowledge against all odds, our modernity, making knowledge the matter of basic subsistence, makes an exception into a commonly shared characteristic, in the process losing the elementary configuration of subtlety and intimacy as access points to anything worth knowing.

The only available solution, as it seems, and in the end this is *fin de siècle*'s message of reconciliation and immense possibility, is to live the precarious coherence of the ruin—of being a "poet in ruins", as Bennett in fact describes Hofmannsthal. The poet himself notes that "All of my pieces adapted from antiquity have to do with the dissolution of the concept of individuality."[46] He takes such a dissolution as the intellectual heritage of the European civilization that eventually leads to "the ego as a manifestation of forces, in its passive as well as in its active aspect, which are synthesized [in a] purified conception of personality, as a glowing center of force and love."[47] In the *Lord Chandos Letter*, this is expressed through a new type of materiality—"a kind of feverish thinking, but thinking in matter [*in einem Material*] more direct, fluid, and glowing than words"[48]—a theme we are more than familiar with by now.

This theme was taken up by another literary figure of *fin-de-siècle* Vienna. Peter Altenberg also set out to disentangle the textures of individuality, drawing a distinction between the 'only' and the 'first':

> The "*Only*" being is worthless, fate's impoverished game with an individual. The "*First*" being is everything! For he has a mission, he is a leader, he knows that the entire humanity follows him! . . . In *all* of humanity, buried deeply, there lies a sensitive, sad, idealistically dreaming poet . . . The poet is never the "*Only*". For then he would be worthless, an abomination of the soul [*Seelen-Freak*]! He is the "*First*" . . . *True* individuality is being ahead of time and alone, what later *all*, *all* must become![49]

The poet as the pinnacle of modern identity leads by quiet example. He is able to create a new range of encounters through his art, but does not become an alienated, detached, and cynical presence.

It is by immersing oneself in namelessness, in a lower state of definition, that another dimension of space eventually opens up. Resonance, given its plasticity and natural dispersion, is the most effective delivery system for this kind of more direct and fluid materiality. Through sonority and rhythm, sound becomes the most attuned form of accelerated duration.[50] In this, sound delivers something that escapes other arts—dissolution of contrast between the space in which one is contained and the one that is actively constructed.

***

The attraction to what we can now describe as the acoustic logic by the generation of artists and intellectuals at the turn of the century is also evident in another Viennese writer, Hermann Broch. In his collection of short stories published under the title of *Die Schuldlosen* (*The Guiltless*),[51] we find (among others) the following titles: "Parable of the Voice", "Voices 1913", "Voices 1923", and "Voices 1933." The theme is the interplay of sound, art, anonymity, and individuality. "Voices 1913", written in verse, presents an assessment and the flavor of European civilization on the threshold of war. Broch incorporated the latest musical trends into the chain of evidence:

> menacing, alluring dissonance/ The triad becomes intolerable and absurd,/ a tradition no longer possible to live in.[52]

"The Methodically Constructed" (1917) begins in the following manner:

> Every work of art must have an exemplary content and must in its uniqueness reveal the unity and universality of all happening: this is true

> of music, especially of music, and in emulation of music it should be possible to construct a work of narrative art in deliberate counterpoint.[53]

The story investigates the innocuous life of a mathematics teacher who revels in the grayness of practical nonexistence, begging the question, "Can such a minimum of personality, such a non-self, be made into an object of human interest? Might one not just as well develop the history of some dead thing, of a shovel, for instance?"[54] The protagonist, indeed, seems to possess a personality of a thing without falling into a state of the inorganic completely. He thus *inhabits* this in-between realm of suspended internment in himself, but with enough endurance that only ambiguity provides.

Broch is one of those writers in whom the consciousness of his generation is heard particularly loudly. In fact, this kind of echo is strong enough for Musil to level a charge of creative larceny against Broch.[55] But it is not only Musil but also Nietzsche and Hofmannsthal who are present behind the scene, given how, according to Broch, Hofmannsthal is characterized by the suppression of the 'I-element.'[56] And Broch carries this message of suppression and erasure of individuality enclosed in names (whether as personality or a phenomenon) exemplarily in his later work, such as *Sailing before a Light Breeze* (1931), in which the 'voice-destinies' of the characters intermingle in the disarray of a newfound irresponsibility to the standard coordinates of presence. This is why

> There's only one defense: namelessness. People who have lost their names can't be called, no one can call them; thank heavens I've forgotten my name.[57]

The action, suspended in the constantly slipping definition of places and characters, seemingly takes place in a Paris café, where a conversation between a man and a woman is overheard by another patron. Their flimsy reality is bent out of shape by a destiny thrown out of alignment:

> For now it was to be anticipated that his voice, which had been etched in the air, would become entangled with the voices of those two, and he calculated at what point in the air before him this might happen . . . Now we're getting somewhere, the young man thought: our voices are becoming beautifully entangled, our destinies will follow. Why should I let myself get further entangled in the destinies of these two people?[58]

This question misses the point, for we find out in the end that there was perhaps no one at the adjacent table, just like the sounds of the café "were very much like silence."[59] Expectations of visibility, of solidified presence, snare perception in the web of deception, whereas only their dissolution, "only death", "can save us from new entanglements."[60] Hence, the entire narrative is punctuated by the expectation of death, literal and euphemistic expiry. The couple, who, according to the silent observer, could be mother and son but then seem to be only two lovers who are playing out their affair in this kind of paternalistic arrangement, are running away from something, expecting to be confronted, caught, exposed, even murdered for their unknown transgression. But the murder had already taken place, because all three, the three in one, are symbolically deflated, given that they have no name, or at least they have learned how names should be given up: "Haven't they got any names for each other? Can it be that they already know how dangerous names are?"[61]

The narrative thus exemplifies a dissonant or, more correctly still, a ruined union, collecting multiplicity into forms without special identity rights, which, like the notes in Schoenberg's compositions, give up their names for the sake of anonymity of expression, creating harmonies with no prior understanding:

> "Understanding is entanglement," said the young man, "and that is my business. If things had no names, there would be no understanding, but there would also be no disaster."[62]

Nonetheless, it is still necessary to acknowledge that "events are spun between names, and without names the world would stand still",[63] which leads to only one possible solution—a reconditioning of the personal pronoun 'I':

> "You mustn't worry, I don't want you to worry," says the boyish voice pleadingly. "I" . . . What does he mean, I? You be still . . . I am I. Even without a name, I am I.[64]

It is this reformulation and preservation of the empty, silent 'I' that leads Broch to conclude that "now all names disgust me",[65] a very close variation on the theme introduced by Hofmannsthal and Chandos's disgust with words. The culmination point of the narrative should also not surprise us, because it is a kind of merger where, as Hofmannsthal proposed, "drei sind Eins: ein Mensch, ein Ding, ein Traum" (three are One: one man, one thing, one dream).[66] In Broch's story the one man, who dreams or imagines the other two, is also actively engaged with the physical environment of the café as living tissue. And what finally takes over is anticipation of silence:

The instruments are being tuned, thinks the young man, and when all voices are in harmony, the moment of death will be at hand.[67]

***

Silence is the hard-learned lesson of Lord Chandos, who by losing the ability speak comprehensively and logically about anything at all instead learns how to perceive indiscriminately, like a radio receiver. By disposing of his linguistic mannerisms, Chandos accomplishes something akin to Freud's ego-psychology, because "Manners are walls, disguised with mirrors . . . Manners are based on a profound conception of the necessity of isolation, while upholding—deliberately upholding—the illusion of contact."[68] Freud and Schoenberg are also preoccupied with exposing and eventually eliminating manners that cloak the real psychology, art, and culture. And culture needs a diagnostic instrument, something like a stethoscope, acoustically mapping a space where

> There is no solid, definable identity between human reality and the word. There can be none. Only the path of indirectness is open, the paradox of surrender to this language while still relativizing or negating it . . . What Kierkegaard [for instance] knew is also the wisdom of *Der Schwierige* [a play by Hofmannsthal; *The Difficult Man*]: "Silence concealed in silence arouses suspicion . . . But silence concealed in a decided talent for conversation, that, as sure as I live, is silence.[69]

Living this kind of topographical ambiguity is possible only as a form of 'speech impairment' without losing one's voice. Kierkegaard's wisdom referenced earlier is evident in the pseudonym *Johannes de Silentio* that leads off *Fear and Trembling*, at the same time promising and eventually enabling new *conditions for audibility*. This is the intensity of psychoanalysis: to facilitate a new power of perception by using language against itself, breaking it apart and thus unmasking the instability of knowledge that is otherwise unassailable.

8 Schoenberg's Oedipus

Freud and Schoenberg never personally met. We know from Stuckenschmidt's biography that in 1971 when the inventory of the composer's library in Brentwood Park, California, was compiled, there were no psychological or psychoanalytic works among the books.[1] This does not, of course, mean that the two men did not know about each other. Given that they occupied the volatile, highly charged environment of Vienna, such ignorance would have been impossible. Psychoanalysis was a much popularized topic of discussion in the intellectual circles, the press, and the Viennese café life. There were also friends and followers circulating between the two creative microcosms instigated by Freud and Schoenberg.

Two of these loyal, 'worldly', lifelong friends on Schoenberg's side, David Josef Bach and Oskar Adler, delved into the larger intellectual context, including the psychoanalytic one, thanks to the milieu of the university—an opportunity not available to Schoenberg. There was also the famous short analysis that Freud performed on Gustav Mahler in the summer of 1910 in Holland. The entire episode is related by Freud's first biographer, Ernst Jones:

> About this time [1910] Gustav Mahler, the famous composer, was greatly distressed about his relationship to his wife, and Dr. Nepallek, a Viennese psycho-analyst who was a relative of Mahler's wife, advised him to consult Freud . . . So they met in an hotel in Leyden and then spent four hours strolling through the town conducting a sort of psycho-analysis. Although Mahler had had no previous contact with psycho-analysis, Freud said he had never met anyone who seemed to understand it so swiftly . . . In the course of the talk Mahler suddenly said that now he understood why his music had always been prevented from achieving the highest rank through the noblest passages, those inspired by the most profound emotions, being spoilt by the intrusion of some commonplace melody. His father, apparently a brutal person, treated his wife badly, and when Mahler was a young boy there was a specially [sic] painful scene between them. It became quite unbearable to the boy, who rushed away from the house. At that moment, however,

> a hurdy-gurdy in the street was grinding out the popular Viennese air 'Ach, du lieber Augustin'. In Mahler's opinion the conjunction of high tragedy and light amusement was from then on inextricably fixed in his mind, and the one mood inevitably brought the other with it.[2]

A direct connection to psychoanalysis also touched the more immediate environment of Schoenberg's professional activity. Both Alban Berg and Anton von Webern were analyzed at one point; Berg consulted Freud in 1908 in an attempt to cure his asthma, and Webern underwent analysis with Alfred Adler, in 1913, soon after Adler's falling-out with Freud.[3]

Besides this, there was the crucial entanglement of Schoenberg's artistic survival in Mahler's private and creative life. In fact, given the intellectual and artistic encounters Schoenberg experienced in the new world of Mahler and especially the composer's wife, Alma, it is impossible to imagine that in the vast range of discussed topics psychoanalysis would not have been among them. The intensity of the suddenly enlarged existential and intellectual horizon in Schoenberg's as well as Mahler's life, thanks to the young Alma Schindler, can be glimpsed from Stuckenschmidt's description:

> Both Mahler and Schoenberg knew little about the events which were taking place round the turn of the century in the artistic scene. When the young Alma-Maria Schindler and her stepfather Carl Moll entered the circle of musicians a new world was opened to Mahler and soon Schoenberg . . . Alma, the housewife and very gifted artist, had been friendly with painters and painting through her father. A strong youthful love had tied her to the much older Gustav Klimt . . . And so in her house—and this corresponded entirely to the wide cultural horizon of Gustav Mahler's—there were meetings of as many painters, sculptors and architects as there were musicians and opera people. Schoenberg, who had hardly any contact in his youth with art, looked with growing interest into a world which opened his eyes in a new way. His later friendship with the architect Adolf Loos, and the painters Oskar Kokoschka, Max Oppenheimer, and Vassily Kandinsky had their foundations in these years.[4]

To this list may also be added Hermann Bahr, Peter Altenberg, and Karl Kraus. It was especially Kraus who proved to be a long-standing catalyst in Schoenberg's intellectual life. Schoenberg himself contributed early on to Kraus's mostly self-published journal *Die Fackel*, whose constant critical commentary on practically every artistic, intellectual, and cultural matter over almost four decades could be described as the model for what today is known as social media—the culture of incessant commentary. That there was considerable bite to Kraus's aphoristic and critical witticism, as well as the sheer volume and speed of his engagement, can be surmised from the fear Kraus spread in the intellectual circles, prompting Hermann Bahr to sue the critic for defamation already in 1901. Kraus, as is well known, was critical of psychoanalysis, famously proclaiming that it was 'the disease of which it claimed to be the cure', but he was not sparing to many other movements, including the fame of Stefan George, whose translations of Shakespeare—he thought—were imbued with the same 'false tonality' that permeated his poetry. This did not prevent Schoenberg from using George's poetry on more than one occasion, including famously in the last movement of his second string quartet, introducing the mood of the entire epoch with the often quoted "*Ich fühle Luft von anderem Planeten*" (I feel the air from other planets). Nor would it prevent Schoenberg from engaging with psychoanalytic thinking in a more serious manner than just a satirical dismissal.

Of course, what Kraus defended and fought for was language, not language as such but its logic and material—its power as a form of expression, not simply a means of communication. Schoenberg imbibed some of that attitude, perhaps more than is generally acknowledged:

> Schoenberg repeatedly draws the comparison between the idea expressed in words and the idea expressed in tones. This suggests that his thought derives not so much from traditional musical thinking, or even from the philosophical aesthetics of music, as from the philosophy of language.[5]

Therefore, as it stands, the situation in relation to psychoanalysis could be described, given Schoenberg's complicated but always expanding network of relationships with some of the leading figures of Viennese and European modernism, as passive awareness. But evidence that it is more than simply such a casual point of interest comes to us in the shape of a drafted and never published essay that proposed to engage the problem of sound and psychoanalysis directly. Assigned the generic title "Psychoanalysis and Music" in the catalogue of the Schoenberg archives and dated 1932, the text meanders from an ironic stance in relation to the pop version of psychoanalytic practice as fashionable talk, or a weapon in critics' hands, to a serious if brief engagement with the concept of the Oedipus complex and its application to tonal music, especially the familial harmony or, rather, the Oedipal transgression that lies at the roots of its definition. Throughout, Schoenberg displays a surprising awareness of some of the more intricate psychoanalytic concepts—more than we would expect, given the official lacuna of knowledge on the topic, if such can be accurately estimated by the contents of what books are to be found in one's library.

The early 1930s were important in Schoenberg's life. His powerful sense of disaster that was to come resulted in a more and more frantic search for existential meaning, enclosed throughout in the notion of the musical thought or idea, according to which sound should take on a deeper and weightier significance. It was also expressed in perhaps his most draining attempt at designing the existential logic of a universal psychology into acoustics—the never completed opera *Moses und Aron*. This philosophical search was in part a response to the more and more calamitous political situation evident in the increasingly unbearable atmosphere at the Prussian Academy of Arts, where Schoenberg held a prestigious professorship since 1926—Ferruccio Busoni's old post. Although still nominally a Protestant, the worsening political situation prompted him to recover his Jewish roots, especially the religious, philosophical mysticism so characteristic of Judaism. Schoenberg began writing *Moses und Aron* in Berlin. He continued working on it in Barcelona, where he stayed for over a year, being granted a leave of absence due to his worsening asthma. Soon after, Schoenberg was to leave the old continent forever.

Schoenberg's creative encounters always engaged the connection between language as voice and sound as a musical construct. The pinnacle of this encounter came in *Moses und Aron*, not only musically but philosophically, through the story itself—its mythology and form that ultimately recoil into silence:

> Being a drama of non-communication, of the primal resistance of intuitive or revealed insight to verbal and plastic incarnation (the refusal of the word to be made flesh), *Moses and Aaron* is, on one vital place . . . a demonstration of the impossibility of finding an exhaustive accord between language and music, between sensual embodiment and the enormous urgency and purity of intended meaning.[6]

The search for this intended meaning and embodiment takes on a multidimensional reality. The year 1932 saw not only the completion of the second and, as it turns out, the last act of *Moses und Aron* but also a draft of a very curious essay on psychoanalysis and music, in which Schoenberg attempts to overcome old prejudices and actively recover something that can become an analytical tool—one that is applied to the psychological reality and its creative extension. Schoenberg's search has always been about the liberation of a voice whose cadence is as yet unknown—mythology is the beginning of looking at the limit of the as yet unheard. And in any case, as Schoenberg at one point admitted to his pupil Alban Berg, everything he wrote had a certain inner likeness to his own self.

***

The draft of "Psychoanalysis and Music" begins in the most obvious location: "Oedipus frees his native city of Thebes by solving the riddle of the Sphinx",[7] and in one gesture sets up the eventual comparison of dissonance to the mythological creature, towards the end of the manuscript. It is precisely due to Oedipus's ability to solve the riddle, seemingly arriving at a resolution, that the prophecy is fulfilled. In Schoenberg's draft the mere *mythos* of the story merges with the *melos* of acoustic transgression.

The point-form opening of what later will be presented in the shape of a more or less completed and continuously drafted narrative enumerates some of the most important psychoanalytic constructs in one vertical column. Thus, in one breath we find "Slips [Parapraxes], Inferiority Complex, Inhibitions, Dreams, Libido, Compulsion to Worthiness/Prestige, Overcompensating Lofty Accomplishments, Neurosis". All these, directly or indirectly, will be used in the impending discussion. A little further down the page, under point number seventeen, we find the Oedipus complex and another sequence of associations: "Inferiority Complex, Abreaction, Repression, Unconscious-Motive,[8] Dream Life". This chain of conceptual formulations, especially its open-ended presentation leading to unanticipated and unusual connections, is crucial to the practice of psychoanalysis and it will eventually become indispensable to Schoenberg's twelve-tone procedure that originated from repetitions of motifs. "A motive is something that gives rise to a motion", writes Schoenberg in 1917. In fact, in a similar fashion to Freud's famous instinct (*der Trieb*) more correctly translated as the drive, one can also "compare the [musical] motive with a driving force."[9]

The ironic tone to which Schoenberg would like to expose the potential reader, at least at the beginning, indicated at one point by the description of the fragment as a 'satirical essay', presents psychoanalysis as a pseudoscience of the type that today we would simply call self-help psychology—something, moreover, that can be used as an effective weapon in the hands of one's enemies—always an active concern on Schoenberg's part. Freud would not be at all averse to such an assessment, given his monumental struggle to keep psychoanalysis free of distorting and corrupting influences. As early as 1914, Freud expressed his dismay at psychological noise making, not insignificantly using a musical metaphor:

> The truth is that these people have picked out a few cultural overtones from the symphony of life and have once more failed to hear the mighty and primordial melody of the instincts. (SE 14: 62)

Schoenberg is more than serious about hearing such a melody through his compositions and his intellectual engagement with various themes—the latter part a much more laborious undertaking. Thus, the subsequent section of the proposed essay, which in its German transcription bears the title "Composers Broaden the Horizons of Your Knowledge", loses much of its ironic tone, instead engaging much more systematically with the theme of psychoanalysis and the question of music, in particular harmony. This analytical encounter between music and psychoanalysis is launched in a fragment that sounds like something out of Nietzsche's *Gay Science*:

> I can only highly recommend to every musician to practice psychoanalysis. He can do this without concern, even when he is made fun of,

> because of it. Art is cheerful [*heiter*]: why shouldn't the artist cheer-up [*erheitern*] and say to himself that in itself the reverence that proves the stupidity of a good thing cannot be harmful, otherwise all the good things would have been harmed a long time ago. And: what damage is it to him, the harm of a good thing? Art is cheerful and maliciousness is also a good thing. Even if psychoanalysis does not prove of much use to one's own self, one can with its help still damage the competitors very easily. And that is worthwhile! ("Psychoanalyse und Musik")

This fragment is still laced with cynicism, but there are also some interesting clues as to the particular shape sound must acquire to survive the imposed constraints of historical tradition. In other words, tones must find an escape route from the 'iron cage' of standard harmonic identity. This type of acoustics is to become cheerful not in its mood but in its technique, which offers the possibility of play with the structure of sound itself.

The longest, continuous section of the draft begins with the question of biography and history, of dealing with the present through the past. Heritage and tradition, important to both Freud and Schoenberg, increasingly appear to be a form of deception:

> To live and to let live: one has to encourage one's biographers. Without the description of life, maybe one has not lived at all; that's why a bad or even an unfavourable description is still better than none at all. But when I have nothing to offer to my biographer, he has also nothing for me. In general, artists should consider their biographies in as broad a manner as possible. They are so well predisposed to it, in any case, that they can not go astray. I have always observed that the mediocre know much better how the brilliant would have to behave, that is why I am often astounded how badly they let themselves be led . . . A far-seeing artist will therefore not let the opportunity slip by to make his life worthy of a story, but he will also be much more laborious in making his work open to interpretation, much interpretation. He will perceive that to the many books must be added: "N. N. and his most intimate secrets, which he himself would not entrust to friends" the countless works about the theme of "most intimate secrets, which he has not dared to admit to himself".
>
> And who should write these, if one does not leave any reasons behind for him? Domain over domain of reasons—soon to come into fruition. But encourage your biographers and you can still experience your belated fame. Vivisection can as much advertise the one on whom the vivisection is performed, as the one who carries it out. ("Psychoanalyse und Musik")

It seems that Schoenberg is learning to think psychoanalytically on the fly. And indeed, after this Strindbergian moment (as we will remember,

Strindberg published two sets of essays, entitled *Vivisections*, in 1877 and 1894, and was an influential figure in Schoenberg's creative life as late as *Moses und Aron*), the composer takes on the theme of psychoanalysis directly, progressively narrowing the field of thinking that led from biography as a symptom of betrayed identity to a system of logic that is enclosed in the structure of sound.

One must begin any new inquiry with a statement of ignorance:

> I do not understand much about psychoanalysis or in other words just as much as those who talk about it. There is no reason then that this should keep me from writing the desired article on the "Psychoanalysis of Music". ("Psychoanalyse und Musik")

Knowledge, as we have heard already, kills action—hence ignorance must be present as the catalyst behind any serious analytical engagement:

> But here one should guard against one error: the problems that underpin such fame [the fame one achieves through a constructed biography] should not steer the author off course [*beisteuren*] since he has to remain beyond reproach: otherwise the art of autopsy in the discovery of problems could not actualize itself! Problems which the author could offer himself must remain unreported: the law of offer and inquiry! No inquiry about the offered problems . . . Inquiry does not like to let itself be outdone by an offer. ("Psychoanalyse und Musik")

If we assume that Schoenberg indeed had not read Freud,[10] or any other psychoanalytic literature (and we have no reason to think otherwise) then the foregoing observation is all the more intriguing. It sounds like Freud's hesitation to reveal himself in the course of dream analysis, always stopping short of completely dissecting his own self. A complete vivisection of one's biographical makeup would interfere with the science of psychoanalysis, because it is not simply remembering that counts (speaking psychoanalytically) but a memory that comes on cue, at the correct time, and in the right proportion so that through the act of remembrance one can also dispense with its burden. And it is indeed interesting that in his caution not to offer too much of oneself to biographers, Schoenberg is really expressing the problem of how to circumvent a standardized identity that destroys the conditions of audibility—something he will mitigate through the kind of embodiment that his sound will progressively assume.

***

Schoenberg's monodrama *Erwartung* was from the start a simulated, psychological vivisection. Freud made such an approach an essential component of his psychology, not only by comparing the psychoanalyst to a surgeon in his more mature writings on technique, but already at the beginning, when considering dreams: "The task which was imposed on me in the dream of carrying out a dissection of *my own body* was thus my *self-analysis*" (SE 5: 454).

The act of slicing through experience is not simply a metaphor but a new perceptive arrangement that combines sensuality and instrumentality. Such a crossover creates a new analytical tool, something like a stethoscope that combines the senses of touch, hearing, and sight, into a multidimensional encounter with cultural presence. Schoenberg is as tactile, through his many artisan projects,[11] as he is aural. And lest we forget, Freud also began with the touch. The application of his hand to the patient's forehead triggered the psychoanalytic procedure that was only then translated into the auditory key. Such translation of touch into sound (similar to any musical instrument) amplifies the invisible organic logic of the human form into layers of psychological and aesthetic textures.

As we learned from Freud and he from the symptom-infrastructure of his patients, the certainty of knowledge builds a wall of resistance because it interrupts questioning, often making it altogether impossible. It is this kind of knowledge in the shape of tonality that Schoenberg reacted against. On this most elementary level there can be no closer kinship between psychoanalysis and acoustics. The artistic personality, whose blueprints were already present in the psychological apparatus of the ego, is indeed very close to the structure of the psyche as such. This connection does not escape Schoenberg, especially because it is often exploited for the wrong reasons:

> What is more intimate, as to say to oneself: as unintelligible and incontrollable as the psyche [*die Seele*] is only the worst and first of all the best music. And motivating factors. Ach, which motivating factors are given to music! How one can engross oneself in them! How one could lose oneself in them!
>
> A musician, and in general an artist, of course on various levels is a genius. At least that! Who then should possess the noteworthy soul [*die Seele*], if not the little genius who can afford the best of the best. It is then the case that to be without ponderous psyche is unthinkable and no expression of genius has the right to remove itself from the research methods of psychology [*Seelenkunde*]. Science above everything else! ("Psychoanalyse und Musik")

There can be no compatibility between the psyche and sound when biography usurps all past and present for its own aggressive purposes. Indirectly, Schoenberg is also reacting against the romantic sound that renders everything disproportionately emotive. What needs to take place, musically speaking, is not prefabricated sound but acoustics based on liberated perception—something that could provide access to the logic of universal relations between things and people enclosed in sound.

That is why Schoenberg dispenses with style, just like the psychoanalytic patient must dispose of the symptom. A symptom presents a simple solution to the intricacies of expression. In a similar fashion to words in Hofmannsthal's reassessment of language, the aesthetics of style stands in for reality. It is precisely tonality that had achieved the status of such symptomatic presence, demanding constant acknowledgment as the only key to the complete sonic environment. The only way is to discard the acoustic biography tonality desires to impose on the texture of sound and the personality of the composer. In his 1934 essay *The Problems of Harmony*, Schoenberg writes,

> We can say that in the development of art, it must always be as it is in Spring! One does what is necessary, though it cause somebody else pain; one does what the situation demands, unconcerned about the approval or disapproval of others. And the cause of music demands, as the history of art-battles shows, that the secret of the sounding tone be always pursued anew. The development of music is more dependent than any other art upon the development of its technique.[12]

Thus emerges another one in a series of correspondences between Schoenberg's art of the tone and psychoanalysis, for the art of psychological interpretation (*Deutungskunst* in Freud's formulation) relies, more than any other art or science, on constant development and calibration of its technique. Both psychoanalysis and the analysis of the tone, as this is what sound eventually becomes in Schoenberg's hands, depend on the *acoustic unconscious*:

> This morning Görgi asked me why ornaments are so rare in modern music. I could not give him a satisfying answer. That people suddenly or gradually lost the taste for them—which is certainly true, at least in part—is, at best, a symptom; but it is probably nothing more than the formulation (for and by the ignorant) of the true (hidden) cause.[13]

That 'true' hidden cause is an inner compulsion, a perceptive state of emergency: "It would not be an exaggeration to call early atonality . . . a state of emergency in the precise sense that a state of emergency is the opposite of a state of affairs in which the law prevails. Yet the emancipation of dissonance . . . was not at all the mere abolition of an old law and the introduction of a new one."[14] In fact, it was closer to a state of exception that does not abolish the law but works from within its confines, rendering it *inoperative*, deactivating its networks, where "that which is deactivated . . . is not annulled, but conserved and held onto for its fulfilment."[15] This is why Schoenberg always insisted that his method of composition was simply the logical conclusion and fulfillment of the acoustic, unconscious core of tonal harmony.

***

> I am not a scientist, therefore I may proceed practically.
>
> Let's begin then with practicality.
>
> Safety first [English in the original]: what does a man wish to know from another most of all, if not whether he is dealing with a friend or an enemy? If his opposition means everything that he says in such a friendly manner, or if he is hiding and hence promotes evil thoughts [?]
>
> Let us take one practical example. A composer wants to sell his work to a publisher, and in the process does everything to do so, is sweet, proud and confident, optimistic and business-savvy; what else would there be for the publisher to find out through psychology [*Seelenkunde*], if not the dammed composer is trying to pull something over his head?
>
> How simple! The composer merely plays a theme from his work and the badly educated in analytical technique publisher follows him according to all the rules of science, and as long as the composer had in reality played conscientiously what is his own and not borrowed from a different source, i.e., stolen, then the publisher knows right away what he is all about and can protect himself, i.e., he offers the same disadvantageous contract that he would have given without psychological analysis, but this time around supported by this science. This is comparable to an insurance company, which thanks to risk-analysis is able to demand more than the mere business competence dictates. It is precisely here that the necessity and practical application of analysis to the motivating factors of music is exposed as far as the publisher is concerned. But I must admit that it is more important, for as far as I am a musician, I am driven by the class-consciousness and interests of musicians to show the advantages that this kind of schooling, this kind of analysis offers to composers.
>
> For instance in the battle for art ("Psychoanalyse und Musik").

Art is the vehicle of expression for sound and psyche, and there is just as much repressed musicality in Freud as one finds symptomatic psychology in Schoenberg. In other words, Schoenberg's 'music' is not simply sound but something like a depth acoustics, where the philosophical, cultural, and personal traits are inscribed into resonance. This analytical orientation that eventually explodes into the twelve-tone method is present already in 1909. In a letter to Ferruccio Busoni, we read,

> And thus I come to answer your other question: how much is intentional and how much is instinctive [in my compositions].
>
> My only intension is to have *no* intentions!

> No formal, architectural or other artistic intensions (except perhaps of capturing the mood of a poem), no aesthetic intentions—none of any kind; at most this:
>
> to place nothing inhibiting in the stream of my unconscious sensations. To allow nothing to infiltrate which may be invoked either by intelligence or consciousness.
>
> If you knew how I have developed, you would have no doubts. But I have prepared myself for this question and am thus able to answer it. I knew one would question the naturalness of my intentions, precisely because they are natural. That one would find them formalized for the very reason that I avoid anything formal.
>
> But when one sees how I have developed in stages, how I was long ago approaching a form of expression to which I now adhere freely and unreservedly, one would understand that nothing unorganic [sic], no "schmock aestheticism", is involved, but that *compulsion* has produced these results.[16]

Such 'therapeutic', compulsive predisposition still felt in its raw form as a type of 'stream of consciousness' in sound—through historical and technical elaboration—will eventually be more tightly controlled, eventually harnessed to a precisely calibrated technique. But the ethos of those early, expressionist, atonal days will never disappear. Hence, in a short text entitled very suggestively *A Self-Analysis*, written in 1948, Schoenberg says provocatively, while looking over his shoulder,

> Atonality or dissonance are no yardsticks for evaluation. Superficiality might base its judgments on such qualities . . . I personally do not find that atonality and dissonance are the outstanding features of my works. They certainly offer obstacles to the understanding of what is really my musical subject.[17]

This understanding, expressed in the form of a new and original *sonic plasticity*, a procedure that allows for aural complexity to be manipulated, depends on very similar procedures Freud employs in explaining the dynamic structure of dream-work:

> At the very start I knew that restriction could be achieved by two methods, condensation and juxtaposition.
>
> Before I could master technically the difficulties of condensation and juxtaposition, I was forced by my destiny upon another road . . . I became the first composer in this period to write shorter compositions.

> Soon thereafter I wrote in the extreme short forms. Although I did not dwell very long in this style, it taught me two things: first, to formulate ideas in an aphoristic manner, which did not require continuations out of formal reasons; secondly, to link ideas together without the use of formal connectives, merely by juxtaposition.[18]

This is how Schoenberg establishes new conditions of audibility—a dynamic space indeed very closely related to the way dreams operate in the human psyche. Condensation, displacement, and juxtaposition are its core components.

Thus, Schoenberg sets himself the sober task of letting the process of *wearing away* applied to overstandardized sound, take hold. This kind of subversive attack on factuality lies at the core of his general relationship to his personal and creative environments, whether it is the technical matter of music, the general cultural environment, or personal affairs that must be squared off with the consistently irritating issue of individuality, especially private biography.

After closing the always anxiety-ridden and unpleasant matter of dealing with publishers in his theoretical sketch, Schoenberg begins to seriously consider the potential relevance of psychoanalysis to music. The first step is to undermine the definition of music by dismissing the importance of not only tradition as tonality but also (and somewhat surprisingly) questioning the value of modern artistic movements:

> Today, as in the past, in every art, there have been many 'movements'; modern, non-modern, classical, romantic, objective and many others. Each strives for its own place under the sun and each disciple, whether for his movement, or at least for his own creativity, seeks space, which he preferably wins by repressing other movements from the public's favour. But such repression is learned. Learned? The science takes care of that; that is its affair.
>
> How? Quite simply, as is already evident in the most difficult case.
>
> Let us assume that a composer, the inventor or a follower of the loosened Homophonic-Polyphony [*aufgelockerte Homophonal-Polyphonie*] is somewhat constrained by the still at the time popularized successes but since some time overtaken movement of "Counterpoint of Seized Curvatures" [*Kontrapunktischen Kurven Ablockerung*]. In that case, he can, quite simply, proceed in the following way: he analyzes a 100 percent exemplary instance of such a style. For example, a *toccata*, a *concerto grosso*, or a *partita*, in a word, a form, the meaning of which is unknown to any living being today. But not in the way that musicians have always done: in light of musical motifs, since moreover, music does not have these but according to hidden causes [*Hintergründe*], since, no doubt

> these it must have; something must be hidden behind such an incessant blather and bawling. A hidden cause [*Hintergrund*], and thus is factual surface [Vordergrund] feigned. ("Psychoanalyse und Musik")

The question here is about directionality: how is the world to orient itself, to which elementary structure should it be affixed? As we already know, Schoenberg does not reject the idea of a hidden or underlying cause *per se*. What he questions is the misuse of such an informative depth. The rejection of standardized perspectives is also motivated by an aversion to a dishonest justification of the status quo, in the service of ideological administration of artistic principles. Schoenberg, then, always stands firmly against any artificial propping up of reality (whether musical, cultural, or personal) that cannot be justifiably prolonged and maintained. This is the only logical consequence of a mutually encrypted feeling of empathy (*Einfühlung*) with the unknown motivating factors that remake identity into a reality under decay. The unconscious forces of low definition carry the kind of uncertainty that Schoenberg designed into sound from the very start, suspending expectations for what should be heard, and thus enabling audibility through the always unpredictable turn of the unforeseen pitch.

The eventual need to break free from the tonal boundaries came from the internal logic of the system itself, not from the superficial phenomena of fashion or trend. This internal working out of tensions is the only form of actual movement through constant displacement of meaning—something that is a unique characteristic of the *fin de siècle*. Here, once again, Schoenberg shares the same space with psychoanalysis that endows internal working out of tensions with utmost importance, precisely under the category of analysis. It is perhaps for this reason that Schoenberg drafted the essay that meant to deal with this exact issue—of how one analyzes effectively, genuinely, and with lasting results, and under his direction acoustics becomes a constantly readjusted process of analysis in sound. The premise here is variability and volatility of perceptive environments. In yet another letter to Busoni, dated August 18, 1909, Schoenberg writes,

> It is *impossible* for a person to have only *one* sensation at a time. One has *thousands* simultaneously. And these thousands can no more readily be added together than an apple and a pear. They go their own ways.
>
> And this variegation, this multifariousness, this *illogicality* which our senses demonstrate, the illogicality presented by their interactions, set forth by some mounting rush of blood, by some reaction of the senses or the nerves, this I should like to have in my music.[19]

It might be objected that this statement, because it comes from the early expressionist period of Schoenberg's creative life, cannot withstand the test of time; that the composer soon thereafter grows out of its somewhat 'childish' phase. Nothing could be further from the truth, and, if anything, this sort of strangely strung emotionality works itself into the mature twelve-tone procedure, even if Schoenberg attempts to control it by continually extending the practical algorithm of how this is to be accomplished, in the end, by some accounts undermining the ethos of his entire paradigm by overly restricting its unpredictability. In this, Schoenberg inadvertently follows a path similar to Freud's engagement, whose early, emotionally overcharged observations are eventually engrained in the nonnegotiable *tour de force* of the technique itself—its various elements that must be present and repeated in a strict line of investigative deployment.

Schoenberg's consistent use of psychoanalytic terminology is no mere accident, and he is more than just a leisurely consumer of its rhetoric:

> The hidden cause and the factuality is feigned; and thus is the hidden cause repressed [*verdrängt*]. Through indifferent fussing about all types of highly charged ideas [*jede Art von bindendem Einfall*] the factual surface will lose sight of the fact that the author, fundamentally and deeply, is nothing more than a sentimental romantic who, ideally, would like to bring into expression beautiful, expansive and original melodies and moods. But he has inhibitions, is ashamed to show himself the way he is, and as all the factual surface (the Inferiority Complex without advancing further the formula of ascertainable=substantial) is to be taken less seriously, the more serious [*Ernst*] is the hidden cause, the conscious-unconsciousness. To take the latter according to its own energy means that repression is not merely proven, but effected [*vollzogen*]. Entire, immense, world-shattering movements, did not they constitute not long ago the latest 'devilish' fashion, have been overcome in this way and now, out of currency, are compelled to give the genius of the next moment his spot in the limelight.
>
> Not to displace the not so joyous task towards the end—for let it cost what it will, even nothing, to be optimistically and happily constituted—I want to say right away that I promise myself few advantages from the proof of abreacted [*abreagierter*] feelings, wishes, and the like. ("Psychoanalyse und Musik")

The process of 'abreaction' was already introduced by Freud and Breuer in *Studies on Hysteria* (1895). It takes on the perennially obstinate problem of forgetting, especially the interrupted sequence of natural 'wearing away' of feelings, meanings, and explanations, which then coagulate into the structure of a symptom. Abreaction is the key to the psychoanalytic technique because it is designed to trigger what natural psychological processes should have performed, but could not—a creation of new lines of association, at the expense of one-dimensional perspectives that force themselves as stand-ins for the totality of life.

Thus, Schoenberg rejects the fashionable, pop version of psychoanalysis, not its conceptual framework—something Freud battles against throughout his career as well. From the perspective of the *fin de siècle*, resonance and psychoanalysis arise through the same medium of time-specific logic, interacting and forming an unwieldy alliance—an alchemy of sorts, where everything is being reevaluated, including the classical, Viennese art-music that must eventually be counteracted with a different metallurgy, a new mixture at the core of reality. It is here, where the logic of the *fin de siècle* finds its parameters, living in the resonant environment of its own echo. This process of unmitigated and relentless juxtaposition of events, concepts, biographies, and phenomena forges a new capacity to produce experiences and the spaces that support them.

Given, then, that experience must be broken apart and submitted to unrelenting scrutiny, Schoenberg finally arrives at another psychoanalytic figure that preoccupies him until the end of his theoretical draft:

> Let us take apart here the difficulty.
>
> Who would benefit, after the analysis of the most deeply inspired symphony [*tiefernste Symphonie*] from the proof that its composer is honestly only 'married to his left hand' and that for all the rejection, all sacrifice, against all profit that this attachment, this overcompensated [unreadable word] has caused him, he in turn compensates by having his right hand leave a mark where he is most receptive, and through this ensure that what he most gladly receives will quickly be lost again. Who, of all the people, would not see that such a strained honesty must be abreacted? And more still: who will gladly find out that it had actually taken place? Who is it that allows himself to be robbed, without the necessity to believe in the ideals of others? Who is it that would not sacrifice himself with pleasure for this, if the way to abreaction would not be postponed for him? This is the wasp-nest. Better not put your hand inside! At this boundary even science should halt: it is best not to know anything!
>
> But things look differently with the Oedipus complex. ("Psychoanalyse und Musik")

The spoiled part of this entire psychological exercise is the one that forces personal psychology into a biographical setting and through it an overdetermined individuality. On the contrary, Schoenberg feels the need for the neutrality of the ego—the self as pure plasticity that transforms the impinging forces of unknown origin into a resolution through complexity. In this configuration, the ego fulfills its own aesthetically driven psychology of wear and tear, only to remodel all the disjointed components into a newly juxtaposed plane of encounters, regardless of whether the latter dons the name of dissonance, the 'I', empathy, or the art of building. Schoenberg does eventually arrive at such a conclusion, although hesitatingly, with many stutters in between. Here, then, is the remaining text of the drafted essay:

It is difficult to think that one could still have friends in life, if one wouldn't know that there is such a thing [as the Oedipus complex] and what it means.

It is indeed unlucky when one is not born with something like that, or at least when one had not managed to acquire it, in the earliest years of one's life. Prosperity inserted belatedly is not so good. Luckily, the good organization and wide dissemination of the modern cultural commodities are making sure that it is practically more difficult not to possess the Oedipus complex than to own an automobile. The mere thought of it already works freely: every small problem can be subdued, every knot, every riddle solved instantly. Let us assume that a composer understands the logic of the theory of harmony [*Harmonielehre*] or tonality. But, alas, such an improbable assumption is impractical—let us better say: Assuming that a composer listens to a good theme (listens to, and not hears, because the issue of whether it originates in his memory or is there thanks to his own invention is open): mixed it up a little, smuggled in one or two wrong notes, things like that happen fast; the standard tonal lineage [*abstammungsgemässe*] would become outdated, its creation time-consuming, but still the wrong place for the wrong notes of the accompaniment is not found so quickly: what is the use of all recklessness [*Sorglosigkeit*] and lack of restraint [*Hemmungslosigkeit*], if one nonetheless comes up against a point where it must be decided between Thebes and Corinth, where one is faced with the question: does it already sound adequately revolting, that one would consider it modern, or, even more irritatingly: does it sound finally so revolting, that one would consider it atonal? Did it pay off to suppress the ancestral harmony, to sully all pure relationships, is patricide worth it? Yes: pure relationships! There it is! Through such impure relationships, no harmony can be maintained, every dissonance turns into the Sphinx!

The mere thought about the Oedipus complex solves everything, as we said. A universal cleaning-agent [*Universal-Waschmittel*]: misplaced [*abhanden gekommenes*] ancestral feeling, destroyed family harmony, inversion of all natural relationships, an impostor [*unrechter Mann*] as unlawful spouse [*unrechter Gatte*], patricide: everything solved; the dirty laundry will be snow-white; the plaited tonality laid aside, the much promising heaps of meaningless tones: a sounding sour-cherry turnover [*Weichselzopf*]; the dissonant connection of confused origins, which still simulate one rule:

Wait, wait, wait!: I have promised too much; the idea of the Oedipus complex cannot solve anything here, only entangle, and the only true word in the entire paragraph above is the sour-cherry turnover [*Weichselzopf*] and I have realized this like in a dream only in the wake of [*im Gefolge*] other association-complexes.

But once again: wait, wait, wait! For should things be thus [?]. The idea of the Oedipus complex solves, in that it entangles. Entangled, if

> one does not take the trouble anymore to separate it out, when one lets it remain entangled—a complex, then—is finally as simple as the dismantled. Analysis collects the disparate into complexes.
>
> He, who in his good conscience has the currently popular courage to become conscious of the Oedipus complex, is cured from his worries about the responsibility for the muddled [*durcheinandergeworfene*] family relationships of harmonies [*Verhältnisse der Familie der Harmonien*].
>
> The submediant, unnaturally in love with the subdominant, claims the prestigious tonic as dominant, on the basis of the same, eternal law, through whose violation [*Vergewaltigung*] it [the tonic] became a tyrant [*Usurpator geworden war*].
>
> This elevates creative courage and diminishes responsibility.
>
> A priceless spoils!
>
> Even before further conclusions.
>
> I can then, only well, say:
>
> Composers, expand your horizons, expand the field of your knowledge. ("Psychoanalyse und Musik")

The argumentative currency of this last fragment expresses the entire dynamic range of Schoenberg's thinking in miniature. We are presented with the deliberate exposure of immemorial lines of descent, ones that violate and pervert the laws of harmony, just like Oedipus did at a seminal point of history—even if that history be fictional and used, as often happened, to remove responsibility for a messy reality from one's hands. This is especially so because it is this elementary fiction at the core of its being that creates the possibility of renewal. And it is not only the violation (*Vergewaltigung*) of the eternal laws of harmony that is of significance. What is really at stake is a new type of stability that collects things that naturally do not go together (of which the story of Oedipus is exemplary) into aggregates or complexes that take responsibility for their transgression, like Oedipus eventually did. Ultimately and notwithstanding all his reservations, Schoenberg seems to see the value of the story in the convoluted nature of misguided heroism that leads to action, the only genuine action that finally exacts the price of undermining oneself.

If we take Stuckenschmidt at his word and accept that Schoenberg's limited formal education diminished his knowledge of the standard mythological repertoire[20]—the signifying symbols one learned along with educational heritage—then his engagement with the story of Oedipus, the attempted reworking of its symbolic matter, becomes that much more interesting. Given his highly practical approach, always searching for the economy of expression whether in his music, his language, and life in general, it is hard to imagine Schoenberg would have wasted time on a purely intellectual pursuit, just for the pleasure of abstraction. Quite to the contrary, we may quite safely assume that the draft of the entire essay on psychoanalysis and music was motivated by a practical search for a formula that could further advance his acoustic paradigm, connecting it to the *economy of transgression* of which psychoanalysis had become the most intense and best advertised expression. Of course, as we have seen, Schoenberg struggles against and eventually rejects the fashion that psychoanalysis had become, but he does take seriously the 'energetic' model of the psyche as formulated by Freud.

The attraction of this myth, the same one that resonated with Schoenberg, is found in the trajectories of escape and avoidance. It is precisely because Oedipus made a deliberate effort to escape from the grasps of a prophesy that had him murder the father and marry his mother, by trying to outrun it, abandoning his native city (or at least thinking he had done so) that he in fact fulfilled all the horror of its potential. This twisted logic that confounds rationality and irrationality in a mutually reinforcing metamorphosis of right and wrong, the traditional and the new, is the logic that Schoenberg seeks. It is something imperative to his own creative progression. But the simple choice between right and wrong, traditional and avant-garde, is also false. It is immature or irresponsible transgression, like the empty act of dispensing with tonality for the sake of being modern or atonal. Schoenberg does not espouse either of these positions, and that is why he maintains throughout that neither dissonance nor atonality are good enough labels for his acoustic output, or even music in general. Rather, what is at stake is a development of a *new acoustic environment*, for which the law of constant deviation and variation, so well prognosticated by the myth of Oedipus, is the modus operandi.

It is through the misapplication of choice in its standardized form that dissonance becomes the Sphinx—a monster preventing the composer and the listener alike from penetrating the traditional, aesthetic secret of sound known as music. Asking whether it actually pays off to suppress pure relationships, Schoenberg concludes with what seems like a contradiction:

> Did it pay off to suppress the ancestral harmony, to sully all pure relationships, is patricide worth it? Yes: pure relationships! There it is! Through such impure relationships, no harmony can be maintained, every dissonance turns into the Sphinx! ("Psychoanalyse und Musik")

We might rightly want to ask, which relationship is Schoenberg referring to, where is the boundary between the pure and the impure, between Corinth and Thebes? It is precisely this question that ultimately must be allowed to fall into disuse, blurring the line between pure and sullied, authentic and contaminated.

From Schoenberg's perspective, this myth, by juxtaposing cultural symbols with the dissonant biographical elements, goes to the heart of compositional technique. Dissonance not only watches over access to musical expression, but above all, it is music's ostracized element. The trick, then, unlike the one Oedipus uses, is not to solve its riddle, or even account for the perplexity it seems to pose, but to eliminate it altogether and from the start, to erase it as a category of understanding. Hence Schoenberg's definition of dissonance as only a more remote consonance, or agreement among sounds, a definition already presented in 1911, at the beginning of the *Theory of Harmony*:

> Now if I continue to use the expressions 'consonance' and 'dissonance', even though they are unwarranted, I do so because there are signs that the evolution of harmony will, in a short time, prove the inadequacy of this classification . . . Since I still have to operate with these notions, I will define consonances as the closer, simpler relations to the fundamental tone, dissonances as those that are more remote, more complicated . . . Efforts to make use of the more remote consonances (today called 'dissonances') as artistic means thus led necessarily to many an error, to many a detour. The way of history, as we can see it in that which has actually been selected by practice from the practicable dissonances, hardly leads here to a correct judgment of the real relations.[21]

It is only at this point that sound is able to correct its own musical history, just like psychoanalysis does with the psyche through, among other procedures, the Oedipus complex. All this comes together in Schoenberg's explicitly drawn space of juxtaposition and displacement that does not so much eliminate standard, family relationships of harmony as it works from the base of their nonexistence as the norm, and hence their strategic violation. It is only at this point that the expectation for a different identity takes hold. And in this sense, we should always think of Schoenberg's music as a delivery system for perceptions that are not purely musical. Sound de-scaled in this way amplifies the textures of the new, decentered psychology—the universal, everyman's condition.

The perennial danger here is the 'one size fits all' attitude, where a formula takes on the shine of a universal solution:

> I do not wish to quarrel with honest efforts to discover tentative laws of art. These efforts are necessary. They are necessary, above all, for the aspiring human mind. Our noblest impulse, the impulse to know and recognize, makes it our duty to search. And even a false theory, if only it was found through genuine searching, is for that reason superior to the complacent certainty of those who reject it because they presume to know—to know, although they themselves have not searched! It is indeed our duty to reflect over and over again upon the mysterious origins of the powers of art (*Kunstwirkungen*). And again and again to begin at the beginning; again and again to examine anew for ourselves and attempt to organize anew for ourselves. . . . If art theory could be content with that, if it could be satisfied with the rewards afforded by honest searching, then one could not object to it. But it is more

> ambitious. It is not content to be merely the attempt to find laws; it professes to have found the *eternal* laws.[22]

And yet one must find a way of observing this attitude of a genuine search, which, by its own endowment and character, must not only violate that which seems to be infallible but also never aim to falsely set up new, infallible realms.

There is then a danger in reckless theorizing about what can only be lived, unless one attempts to solve problems through more entanglement, bringing together "the disparate into complexes". The weight of this last formulation is indeed imposing, because it basically explains Schoenberg's mature art, its theory and technique of composition.

The twelve-tone method and its earlier atonal condition attempt to accomplish precisely that. They collect notes that have lost their native environment of a key, into new complexes. To those interned in the norm of tonality, these new relationships seem not merely vagrant but perverted, an abomination of natural sound. A parallel sequence of events to the Oedipus myth suggests itself—topography of arrivals and departures, of loss and acquisition, and the confusion between the native/original and the acquired. This is in part why the story, especially as it is taken over by the psychoanalytic practice, combining elements that are fundamentally at odds with each other, is so attractive to Schoenberg. Simply put, the myth represents in efficient strokes the otherwise hopelessly convoluted economy of transgression. In this, the idea of a complex, taken on the more abstract level of a form, provides a solution even before the puzzle is encountered, arriving at an answer outside the logic of a riddle. This sentiment is again expressed by Schoenberg in the pages of his *Theory of Harmony*, in the preface to the first edition of the book:

> There is a mechanical puzzle that consists of three small metal tubes of different diameters sealed in a glass-covered box. The problem is to get the smaller tubes inside the larger. Now one can try to do it methodically; then it usually takes quite a long time. But it can also be done another way. One can just shake the box at random until the tubes are inside one another. Does that happen by chance? It looks that way, but I don't think so. For an idea lurks behind this method. Namely, that *movement alone can succeed where deliberation fails.*[23]

It is the act of nonacceptance, of refusal to consider the puzzle in its own identity, not to accord it the respect and seriousness of a problem that resolves the complication. Thus the seemingly irresolvable (except by its avoidance) quandary of dissonance is preempted and outmaneuvered by movement that operates according to the natural inclination of phenomena—a procedure that dominates Schoenberg's attitude towards sound and composition from the beginning. But one needs to be continually vigilant

about the attraction of quick fixes, which purify relationships, like someone who becomes conscious of the Oedipus complex, only to be "cured from his worries about the responsibility for the muddled [*durcheinandergeworfene*] family relationships of harmonies [*Verhältnisse der Familie der Harmonien*] ("Psychoanalyse und Musik").
Contrary to such psychological cosmetics, the Oedipus complex taken according to its own valance instigates cross infiltration and eventual internal collapse of perspectives. This process of topographical ruination that reassigns the previously undisputed position of certain tones to new locations should extend as far as possible:

> Merely in order to retain the metaphor of rising and falling intervals, I speak of the *ascending* interval as a fourth *upward* and of the *descending* as a fourth *downward*. (This is only a metaphor, as is our designation of pitches as high and low. Since the tones are literally neither high nor low, we could just as well express this distinction by means of other antitheses: for example, sharp and blunt, short and long, etc.).[24]

The distinction between high and low, just like the artificially introduced contrast between consonance and dissonance, usurps the power and freedom of evolving definitions. It blocks the power of neutral plasticity, the state of highly charged malleability that is always ready for new combination of elements. And indeed this is the final accusation against tonality—a space that elevates the tonic to the position of acoustic dictatorship:

> The submediant, unnaturally in love with the subdominant, claims the prestigious tonic as dominant, on the basis of the same eternal law, through whose violation [*Vergewaltigung*] it [the tonic] became a tyrant [*Usurpator geworden war*]. ("Psychoanalyse und Musik")

The perversion here is the replaying of the Oedipal drama on a different scale, or, in fact, in the musical scale, where tones are combined according to artificially erected familial relationships, acquiring the aura of immutable legitimacy. The elevation of the tonic (the first note of a musical scale that gives a piece of music its key) to the highest indicatory value, where it is able to claim power and exercise control over the entire musical range, is in itself, according to Schoenberg, something close to musical incest. Regularity, correctness of behavior, and restricted interaction constitute distortions of the otherwise much more fluid acoustic universe, like the one already contained in the musical complex known as the six-four chord:

> A unique claim is thus made for the fourth [one of the defining distance-components of the six-four chord]: that its two tones form now a dissonance (when the lower tone is in the bass), now a consonance (when it is 'covered').[25]

The six-four chord, technically speaking already a second inversion of the most fundamental triad, because it rearranges and complicates distances among the family of pitches that make it up, is ambiguous, given how it is perceived on certain occasions as more or less dissonant. The same logic motivates Schoenberg's translation of the Oedipal myth into the technical language of harmony. The submediant or the sixth degree of the musical scale is in love with the subdominant because it shares two of its three tones with the subdominant, allowing for an easy transformation into its other—for instance, from A minor (A-C-E) to F major (F-A-C). This transformation is traditionally known as *Leittonwechsel* (exchange of lead-tone), used extensively in tonal modulation. The point is that Schoenberg extends this type of transformative logic. He sets up an entire system of how this initially weak or unstable point can become a new space that 'music' now inhabits, underpinned by a set of specific rules of how this kind of acoustic morphology is to be accomplished.

The standard musical triad (the most basic relationship of three tones 'stacked' vertically on top of each other) preserves the artificially calibrated distances between pitches, measured out as specific configurations of sound. Schoenberg stretches these distances, displacing the fundamental tone or the root (which is the most direct way for a composition to acquire its tonal identity) from the position of leadership, scrambling all sonic relationships. This is how the accountability to the stringently regulated tonal expectation is now loosened:

> Any setting of a chord that has the root in the bass imitates most closely the acoustical relation of the tone to its overtones [and] once we put another tone in the lowest position, we depart from the natural model.[26]

Schoenberg is certainly not the only composer to have done this, but he is definitely the one who has gone the furthest in this process of wearing away of musical standards, setting into motion the process of ruination applied to tonal expression.

Harmony, as the system that maintains, regulates, and preserves musical distances among tones, works according to the dynamics of repression. The result is a misalignment between the surface aesthetics of sound and the deeply seated acoustic drive. Again, the six-four chord is one of the most vivid instances of this sleight-of-hand, because it embodies "a conflict between its (outward) form, its sound, and its (inner) constitution."[27] And this conflict between the outward form and the inner constitution is the axis on which the psychoanalytic model of the mind also rests. Schoenberg uses almost exactly the same models for sound, concluding that the resolution of further dissonances from the simulated one in the six-four chord will of necessity display and require "a different psychology."[28]

Such psychologically endowed acoustics directly echoes the psyche, at times supporting, and often contradicting and repressing the depth-psychology of inner drives. This acoustic drama is again located in the peculiar circumstances of the six-four chord:

> The six-four chord, whose problem could be felt but not understood, was always treated in a special way. This fact alone would have been enough to give it a unique position . . . But whether its unique position has its origin in convention or in nature, that position is nonetheless quite definite. And since in practice the chord was always regarded as a singular phenomenon, to be used in a special way, since particular events always preceded it and particular events always followed, it has an effect similar to that produced, say, when part of a popular quotation is cited: 'Love conquers . . . '—one fills in what comes before and after. The merest hint at the well-known cause awakens the expectation of the well-known effect. With the first word we understand; we then await a particular continuation. This is how the cliché, the formula works. The six-four chord developed into just that, a permanent formula with the effect of a cliché; one cannot imagine its occurring out of the accustomed context.[29]

The clinching point is the already familiar conclusion: "the conflict in the six-four chord and its tendency toward resolution are not absolutely compulsory; one does not have to give in, for nothing more than overtones support this tendency"[30]—just like standard family relationships are supported by nothing more than social conventions.

What is it, then, that is at work in Schoenberg's cultural, creative, and musical world, something that is also an echo of the *fin de siècle*? The easiest way to describe the logic of his acoustic art is through the mechanics of association that *inverts* normality. In this, by Schoenberg's own pronouncement, "music is comparable to a gas, which is in itself without form but of unlimited extension."[31] This means that the encounter between forces and events is expandable almost to the point of infinity, and that the task of art, especially music, is to capture this moment of the potentially limitless effusion, just like dreams always renew their material through a representation that spills over the previously set boundaries:

> And—without saying that the ultramodern music is really atonal: for it may be perhaps that we simply do not yet know how to explain the tonality, or something corresponding to tonality, in modern music—the analogy with infinity could hardly be made more vivid than through a fluctuating, so to speak, *unending harmony*, through a harmony that does not always carry with it certificate of domicile and passport carefully indicating country of origin and destination.[32]

A homeless, vagrant harmony, then, that had lost its coordinates of subordination—all this is not unlike Oedipus's trajectory of setting out on the road, losing, gaining, and then losing again the hospitality of a domicile and the reassurance of familial homogeneity. And should we really be surprised that Schoenberg writes this already in 1911, an opinion (or more accurately a complex of conceptual formulations) that exists in a more or less complete form without giving us the name of Oedipus and his complex, as is done in the drafted essay on psychoanalysis and music from 1932? Schoenberg, not unlike Freud, always operates through stages of completion, where the maturity of the entire system shows itself in the nascent state of variable emphasis. Hence, we can venture out and propose to those who feel adventurous enough that the *Harmonielehre* is already written psychoanalytically, or at least that it employs a similar image-palette, similar logic—an indication that both Schoenberg and Freud are simply responding in a similar manner to the *historical logic* of their times:

> Napoleon, who installs his relatives and friends on the European thrones. I think that would indeed be enough to explain why one is justified in obeying the will of the fundamental tone: gratefulness to the progenitor and dependence on him. He is Alpha and Omega. That is morally right, so long as no other moral code obtains. Yet, another can indeed prevail! . . . It is superfluous to hark back every time to the ancestors, to show every time with utmost thoroughness the derivation of chords from the fundamental; it is superfluous to prove their lineage every time when it is familiar to everyone and lives in everyone's memory and sense of form . . . And whenever we let our imagination roam, we certainly *do not keep ourselves strictly within boundaries, even though our bodies do have them.*
>
> The sense of form of the present does not demand this exaggerated intelligibility produced by working out the tonality. A piece can also be intelligible to us when the relationship to the fundamental is not treated as basic; it can be intelligible even when the tonality is kept, so to speak, flexible, fluctuating (*schwebend*).[33]

Eventually, the fluctuation at the core of Schoenberg's musical technique will acquire its own parameters, its own evenness of progression and suspension, making it comparable to one of the mainstays of Freud's paradigm, what he calls '*gleichschwebende Aufmerksamkeit*'—the evenly suspended (or evenly fluctuating) attention. And all of this is premised on the activity to which the composer is called in the concluding sentence of the drafted essay on psychoanalysis and music: "Composers, expand your horizons, expand the field of your knowledge." Expansion inadvertently turns knowledge into an inquiry. The constant treatment of reality as though with sandpaper reverberates in the activity of notes and individuals. They must find a different value and justification for their presence, because they are judged by the harshest standard of evaluation there is—by what lies beyond their 'native' comfort:

> What is within reach has its temporary boundaries wherever our nature and the instruments we have invented have their temporary boundaries. What is attainable with the phenomenon outside ourselves, as far as the tone itself is concerned, theoretically speaking, has no boundaries.[34]

Acknowledgments: "Psychoanalyse und Musik" from the Archives of the Arnold Schoenberg Center in Vienna [ca. 1932], T39.23. Used by permission of Belmont Music Publishers. All translations are mine.

9 Interpretation of Dreams
Theory of Harmony

At the turn of the twentieth century a new dimension of presence, requiring that distances and intervals, including the musical ones, were measured by the standards of something previously unheard, came to the fore:

> The tendency of something unheard of to reveal itself to the searcher is just as great as is the tendency of the searcher himself to find something unheard-of. And in this sense every truly great artist is an impressionist: ultrafine reaction to the slightest stimuli reveals to him the unheard-of, the new.[1]

The adoption of the 'unheard-of' element as the base of engagement with art, especially the art of sound as creative event, takes on the shape of an incalculable economy—resonance being its most direct expression:

> In music, along with the content of its smallest components (tone, tone progressions, motive, *Gestalt* [pattern, figure], phrase, etc.), there is an additional means of economy available, the possibility of sounding simultaneously. Perhaps for this reason it says more to everyone than do other arts.[2]

It is such an alternate range of perception that takes hold of Schoenberg, because, as he says on another occasion, the intervals must not be necessarily measured according to the usual arithmetic, where the distance spanning the notes, like C to G, for instance, appears as five increments away in every context.[3] This inverted logic deflects perception into unprecedented arrangements, like in a dream, because "whenever they [dreams] show us two elements close together this guarantees that there is some specially intimate connection" (SE 4: 314).

From the start, then, we are in the kind of space on whose premises identity presents itself as a network of unlikely juxtapositions—the space of Freud's dream-work (*Traumarbeit*) and Schoenberg's acoustic complexes. We encounter here an economy beyond value, something in the range of Freud's *unheimlich*, and all this already in the course of reading a dedication to Gustav Mahler, whom paradoxically Schoenberg considered invaluable, opening the pages of *Theory of Harmony*. It is a dedication that is given and immediately dissolved, the symptom of Schoenberg's peculiar brand of thinking, which he shares with not only Freud but also the entire context of the *fin de siècle*.

The dedication appears in 1911, in the first edition of Schoenberg's *Theory of Harmony*. By the time of the second, 1922 edition, it is substantially modified:

> The sentences which in the first edition accompanied the dedication to Gustav Mahler are now omitted as superfluous. They were words written in profound grief directly after Mahler's death . . . The declaration that 'He was an altogether great man', being unproved [at the time], thereby won very nearly the power and the repute of a prophecy. But this declaration, which indeed was intended to win in order to give, won more than it gave; to him, who has everything, it can give as much but no more.[4]

One cannot but notice a *psychology of dissolutions* that is immediately activated in the pages of the book, the same one that Schoenberg applies to his treatment of sound- matter, and the kind of representation that constitutes the building blocks of the psyche, as Freud expresses it. Ultimately, it takes the form of a new complex, what we might call the *echo-complex* that rewires the basic, material constitution of musical tones. As we might by now suspect, tones are weaved together using various elements and dimensions, not only acoustic frequencies:

> Now, if it is possible to create patterns out of tone colors that are differentiated according to pitch, patterns we call 'melodies', progressions, whose coherence (*Zusammenhang*) evokes an effect analogous to thought processes, then it must also be possible to make such progressions out of the tone colors of the other dimensions . . . That has the appearance of a futuristic fantasy and is probably just that. But it is one which, I firmly believe, will be realized. I firmly believe it is capable of heightening in an unprecedented manner the sensory, intellectual, and spiritual pleasure offered by art. *I firmly believe that it will bring us closer to the illusory stuff of our dreams; that it will expand our relationships to that which seems to us today inanimate* as we give life from our life to that which is temporarily dead for us, but dead only by virtue of the slight connection we have with it.[5]

There is only one type of materiality that can satisfy such a desire, the tone as the most malleable of physical and, more crucially, cultural forces, but not in its current configuration that suppresses its full potential. Under the name of tonality, tones are allowed to move only in certain directions. Liberated from this overarching identity and catapulted into a network of forces, flows, and intensities, the tone acquires the profile of a new materiality.

The *fin de siècle* deploys a long series of such spatial reconfigurations. Not only the flashy choreography of the avant-garde movements but also the complete reassembly of perception are its effects. This much has been already observed, no better perhaps than by Henri Lefebvre:

> The fact is that around 1910 a certain space was shattered. It was the space of common sense, of knowledge (*savoir*), of social practice, of political power, a space hitherto enshrined in everyday discourse, just as in abstract thought, as the environment of and channel for communications; the space, too, of classical perspective and geometry, developed from the Renaissance onwards on the basis of the Greek tradition (Euclid, logic) and bodied forth in Western art and philosophy, as in the form of the city and town. Such were the shocks and onslaughts suffered by this space that today it retains but a feeble pedagogical reality, and then only with great difficulty.[6]

The shudder that reverberated through political, social, and artistic realms triggered a type of redirected attention, displacing communication from its simple confinement *in* space. Instead, space itself in the form of the psyche, the body, or the tone became a technique of analysis. It is in this form that 'analysis' and 'space' were fused together into complexes of materially redesigned assemblages, splicing distance and proximity, stability and crisis, identity and anonymity, and so on. All these became so many echoes of the formative logic that lies at the bottom of the *fin de siècle*. It is the logic of *plastic manipulation* that employs the entire sensory spectrum, as well as the various technologies derived from it, in an unrelenting pursuit of ever more viscous material, which could yet again be recombined, just like "the construction of collective and composite figures is one of the chief methods by which condensation operates in dreams" (SE 4: 293).

"You know the name you were given, you do not know the name you have", wrote the Portuguese novelist José Saramago decades later, when the twentieth century was already in full swing.[7] Schoenberg ascribes a special status to such unknown and unformed names. Sound responds by releasing the logic of combinations for which the blueprint is the far-ranging, multimedia architectonics of the ear. It is a system that "reflects the way in which our ear draws analogies (*kombiniert*): it connects like things, it sets widely separated events adjacent to one another and adjacent events over one another"[8]—precisely the type of coordination Freud ascribes to the field of dreams that to him are a new range of the factual.

***

Remembering Schoenberg's conclusion that a reformulated nature of the tone could finally bring us closer to the "illusory stuff of our dreams", we can at last bring the theory of dreams and theory of harmony together. From the start, Freud is exposing and popularizing the migratory nature of perception. The essential point is that analysis becomes a mass medium of communication, infiltrating subsequent and ever extensive realms of culture. This is the case especially with art that assumes the role of a code into which universal relations are translated.

The tone and the dream, then, are two phenomena united by the similarity of their operational principles:

> In the first place, dreams take into account a general way the connection which undeniably exists between all the portions of the dream-thoughts by combining the whole material into a single situation or event. They reproduce *logical connection* by *simultaneity* in time. (SE 4: 314)

This is how the psychological event and the musical moment find their ground—by emancipating the near simultaneity of communication. The performance of this type of operation depends on the interaction of several processes that make up the system Freud describes simply as the dream-work (*Traumarbeit*)—the components in question are displacement (*Verschiebung*), condensation (*Verdichtung*), and distortion (*Entstellung*).

Dreams and tones are nomadic figures that need their own source of energy. The latter is generated by exploiting tensions that arise from combining, metamorphosing, and juxtaposing experiences, thoughts, and desires into altogether unprecedented assemblages:

> These usually emerge as a complex of thoughts and memories of the most intricate possible structure, with all the attributes of the trains of thought familiar to us from waking life. They are not infrequently trains of thought starting out from more than one centre, though having many points of contact. Each train of thought is almost invariably accompanied by its contradictory counterpart, linked with it by antithetical association. (SE 4: 311–312)

It is also in the nature of every chord to lead away, to go astray from the center of the fundamental tone and its imposed key, installed as the overseer of 'correct' but otherwise arbitrary harmonic progression:

> After all, if one thinks about tonality [one can see that] everything, with the exception of the first degree, [the tonic], is so to speak passing, or at least going—everything is in motion.[9]

Schoenberg makes it a point of emphasis that one does not need any special or elaborate means to escape the musical key because every complex of sounds that follows the initial statement of tonality is already a sonic saboteur, inhabiting the remnants of another center, another complex. The conclusion then becomes clear: the chords that make up present musical language also carry within a freeing instability that always pushes the system outside of its blindly maintained frames of reference.

Theory of Harmony should be read as a work of modern psychology and a model of aesthetic perception, not simply as a technical manual of composition, because in Schoenberg's understanding tonality, just like Freud's ego-psychology, is fundamentally at odds with itself, caught up in a series of self-contradictions whose framework generates productive instability. This psychological language is not simply an occasional metaphor but a staple configuration where drives, the sonic unconscious, repression, and dreams eventually condense into acoustic expressions.

The mode of operation here is the power to decompose values, judgments, and boundaries. This is how dreams, too, combine experiences, especially their leftovers, into variable arrangements. At first, what is of most essential value to dreams seems to be of no value at all to reality:

> The way in which dreams treat the category of contraries and contradictions is highly remarkable. It is simply disregarded. 'No' seems not to exist so far as dreams are concerned. They show a particular preference for combining contraries into unity or for representing them as one and the same thing. (SE 4: 318)

The preference for the nonessential materials, insignificant encounters, inconsequential experiences, illogical words, images, and sounds creates a space without up or down, inside or outside, where dissonance and contradiction are eliminated, because one does not acknowledge the prohibition in the first place. What remains is a continually redrawn field of contrasts.

We can perhaps now understand more clearly Schoenberg's illustrative use of the Oedipus complex. Dissonance in the functionality of the tonal system is akin to a guilty conscience that shrieks with terror before a transgression. What we find in the nature of the tone is the same translation mechanism that operates in dreams, and this would be a good time to remember that according to Freud dream interpretation is the archetype model for the workings of the psychic apparatus and culture in general. Meaning, such as there is, is located in transition zones between the latent and the manifest content of phenomena, where the turbulent, heretical information thrives in a masked form. This dynamic is repeated in sound, through what Schoenberg refers to as vagrant chords:

> Little is actually said whenever one shows where the chord comes from. Because it can come from everywhere. What is essential for us is its function, and that is revealed when we know the possibilities the chord affords. Why single out these vagrant chords and insist that they be

> traced back at all cost to a key? Later, the pupil will best take all these vagrant chords for what they really are, without tracing them back to a key or a degree: homeless phenomena, unbelievably adaptable and unbelievably lacking in independence; spies, who ferret out weaknesses and use them to cause confusion; turncoats, to whom abandonment of their individuality is an end in itself; agitators in every respect, but above all: most amusing fellows.[10]

Sound is a vagrant reality, in a state of autoanalytic frenzy. Eventually, Schoenberg will dispense with the idea of vagrant or nonharmonic tones entirely, for such descriptions can make sense only in the strictures of the tonal system, a system whose constraints one is now able to escape: "*There are no non-harmonic tones, for harmony means tones sounding together* (*Zusammenklang*)."[11] As always, the removal of tonal censorship, the act of stepping outside its boundaries, should be motivated by the search for new parameters of cultural presence, because the artist "is merely the instrument of a will hidden from him, of instinct [*der Trieb*] of his unconscious . . . He feels only the instinctual compulsion, which he must obey."[12] Thus, identity is split between two forces, a division matching the psychoanalytic paradigm. Not only that, but in a similar manner to Freud's modeling of the psyche, these impulses compete and struggle against each other for the possession of absolute rights and the unchallenged power to define all relationships of distance and proximity.

The Freudian moment in these concluding pages of Schoenberg's most sustained theoretical exploration of sound and culture is elevated to such a pitch that even the translator of the work, Ray E. Carter, calls for a speculation about a possible connection.[13] Everything points towards the conclusion that Schoenberg, in spite of his official ignorance, was much more familiar and intimate with Freudian theory than most commentators cared to admit. But this realization is also a misnomer, for it misses the convergence of Schoenberg's music with psychoanalysis that lies in a much deeper connection. For we may ask, why is it that Schoenberg feels the necessity to use psychological/psychoanalytic language at all, at this still relatively early stage of his development? And the most likely inference is that he lacks a mode of expression to stabilize and expose the logic that permeates practically all levels of the *fin-de-siècle* culture. This is the logic of space in *dispersal*, a process I have already presented under the paradigm of inhabited ruins and the aesthetics of disappearance. Once again, we are placed in a reality where

> Indeed, distance and proximity are no longer mutually exclusive and the space or scene of the unconscious must be rethought to take account of this non-exclusivity. The 'I' that remembers, like the 'I' of the dreamer, finds itself scattered throughout the dream despite its apparent distance from it. To be both nearby and far away at one and the same time is to be subject to the effects of a certain *dispersion*. And indeed, Freud

> insists that the dream can only be understood as a form in which the I abandons itself to such dispersion.[14]

Schoenberg uses resonance to set up this type of controlled dispersion of identity that captures the logic of presence, circa 1900.

But the unconscious, a concept with a long tradition in German philosophical thought, is also a crutch. Its form must be reconfigured to escape from the 'natural' condition of hearing impairment. Freud is always hard at work to pull himself away from its purified formula:

> It is not without intention that I speak of 'our' unconscious. For what I thus describe is not the same as the unconscious of the philosophers or even the unconscious of Lipps. By them the term is used merely to indicate a contrast with the conscious. (SE 5: 614)

***

Just as Freud sets out to prove that one does not simply dream but rather *lives* on the plane of invention through dreaming, so too Schoenberg insists that "one does not harmonize, one *invents* with harmony."[15] And what one invents is nothing less than new nature, new individuality: "a new sound is a symbol, discovered involuntarily, a symbol proclaiming the new man who asserts his individuality."[16] This is the case not only because sound through its identity vested in the structure of the chord has not been described adequately up till then but also because tones themselves are animated, organic structures: "the tone lives and seeks to propagate itself."[17] Such organic sound can be manipulated like a physical body, even dissected. It is like work "in dissection performed by young doctors on cadavers: if they cut too deeply, nobody's hurt."[18] And even though Schoenberg here refers to the exercises to be performed by the student of harmony on dead musical fragments that are far from being fully fledged compositions, this general notion of analysis also applies to the work of the composer. Again, Schoenberg says this directly: "our relation to this prototype [the tone] is that of the analyst, of the seeker; in imitating it we discover more or fewer of its truths."[19] The role of the analyst, like the one emphasized by psychoanalysis, is not simply to shatter or break apart (in the strictest meaning of analysis) but to provide a perceptive ligament that could connect divergent impulses, *transferring* tensions between various dimensions.

Towards the end of *Theory of Harmony*, in the section entitled "The Chromatic Scale as a Basis for Tonality", Schoenberg responds directly to a critic, a certain Dr. Robert Neumann, having no qualms about the psychological assumption that underlies his entire musical universe:

> Most essential is the following *psychological assumption*: The development of the harmonic resources is explained primarily through the

> conscious or unconscious imitation of a prototype; every imitation so produced can then itself become a prototype that can in turn be imitated.[20]

This explanation enables an interval-like, mimetic orientation applied to a range of occurrences, and it goes beyond the strictly speaking 'musical' reality, if only because very similar mechanics are found in related aesthetic and intellectual phenomena. Tonality overregulates distances, but in itself it is only a symptom of something else. The objectionable part of any system, including the tonal one in the history of Western sound, is its indiscriminate use, at one point justified simply because one follows the rules, not because it is an internal necessity. Such a mode of engagement with sound-matter represses the aesthetics of invention, benefiting the externally imposed standards of how one is allowed to perceive and create. But the inner impulse in tonality also tells a larger story. The use of a tonal key is a decision that should be made purely for creative reasons, although, according to Schoenberg, its use at the beginning of the twentieth century increasingly missed the point of contact with the world at large, given how it did not have the power to express the shape of modern life. In other words, standard harmony did not correspond to the prevailing logic of how relationships as such were being constructed—according to the logic of plasticity. Since 1900 it was this plastic element that was increasingly marking the perceptive and communicative threshold of phenomena—in itself an expression of the type of pan-materiality espoused by Adolf Loos, for instance, both in its character and use:

> One must not be ashamed of satisfying the practical requirements of the material, and may confess it openly, without adorning it, without glorifying it. But then one will not always do the opposite of what is right. Those who on the one hand simplify (by suppressing sense and preserving only its shell, the formula) complicate on the other hand (by 'artistically' embellishing the shell). Thus they fail in everything that matters: because instead of sense they always give the form, the formula. *Their simplicity and their complexity are incorrectly related to the content.* They are simple when one must be complex; they are intricate when one may be straightforward. I have known this for a long time in my own rather narrow field, but in another I had to learn it from a rough slap in the face, which I deserved. That slap made clear to me how much our taste in almost all fields is corrupted by the 'decorators' (*Ornamentierer*) (as Adolf Loos calls them) disguised as simplifiers.[21]

This is how tone-psychology and ego-psychology are entwined, related, and mixed up in the same reality of correctly calibrated simplicity and complexity, inhabiting one space of logical relations at two different levels:

> One and one only of these logical relations is very highly favoured by the mechanism of dream-formation; namely, the relation of similarity,

> consonance or approximation—the relation of 'just as'. This relation, unlike any other, is capable of being represented in dreams in a variety of ways. Parallels or instances of 'just as' inherent in the material of the dream-thoughts constitute the first foundations for the construction of a dream . . . Similarity, consonance, the possession of common attributes—all these are represented in dreams by unification, which may either be present already in the material of the dream-thoughts or may be freshly constructed. The first of these possibilities may be described as 'identification' and the second as 'composition.' (SE 4: 319–320)

The fact that Schoenberg says something very similar about sound, not only treating dissonance as simply a more distant agreement, but always placing the acoustic part beyond the strictly musical form, is that much more remarkable. This suspension of tonality and its eventual replacement inadvertently create a new form of expectation, taking the shape of exception, where "the exception is not a mere exclusion, but an *inclusive exclusion*, an *ex-ceptio* in the literal sense of the term: a seizing of the outside."[22]

We are thus continually discovering a psychology where one does not exercise direct control, because the tone does not respond to its representation from within the system, but to a hidden current that touches all reality:

> Instead of seeing that both [the chord and the melody] serve only one purpose, that is, penetration into what is given in nature, we take now the one, now the other to be the essence of music; whereas, in reality, this essence is not even the third [possibility] but rather some fourth [possibility] that cannot be more closely examined here.[23]

The translator of *Theory of Harmony* summarizes the options, concluding quickly that we are left with a mysterious emptiness, because, according to Schoenberg, the essence of music is not found in the chord, or counterpoint, or even harmony in its most advanced form, eventually derived from the acoustic shape known as the twelve-tone row. The logic of sound is perhaps found in what Schoenberg refers to as the musical idea—but here we would have to start again, attempting to define its parameters—precisely what the composer does throughout his creative life. But there is an indication of what this fourth or 'nth' possibility after the exhaustion of every other paradigm might be—and it comes from the cultural and historical context itself. It is something like the *threshold of fragmentation* (*Schwelle der Zerkleinerung*).[24] When phenomena, including sound, fall below this threshold, they fragment into a cloud of diffused incidents. On the other hand, being too far removed from it, as is the case with standard tonal harmony, they take on the false gravitas of being directed, even if by inaction. In other words, such a threshold constitutes the logical and material core of perception—a new type of operational space practically all events passed through in the first few decades of the last century—a space where cultural identity as a certain topography of historical consciousness comes together and encounters cultural perception not simply as a mute registry of information but as the *landscape for action.*

Is not Schoenberg's sound persistently approaching this type of a threshold, attempting to inhabit its inhospitable dimensions—a thin line indeed, beyond which there is nothing left? It is the sort of dimension that brings together simplicity and complexity, distance and proximity, stability and dissipation. Sound is a ready-made complex of all these moments, of raptures brushing lightly against the sense of a potential disaster because now the boundaries that separated one reality from another have been practically expunged. And just as Schoenberg finally concludes that there is no such a thing as nonharmonic tones, Freud too from the very beginning realized that there are really no dreams as such, but only configurations of expressive intensity that coordinate our 'disheveled' world.

But how does one hear such voluminous streams of information, when these operate on a frequency that might not be immediately accessible? The answer is now predictable: by calibrating and fine-tuning a special organ of perception. It is the ego in the psychoanalytic model that becomes such an organ, expressive of the historical moment in which it was created and whose echo it still constitutes: "we regard consciousness as a sense organ which perceives data that arise elsewhere" (SE 4: 144). Consciousness is the receptor that stands in between the latent and the manifest organization of the entire store of messages, perceptions, and occurrences. Depending on their charge they are treated through variable sequencing and fluctuating amortization. In this way, we acquire an organ of distance management, something like an unconscious ear:

> What has been attained is the almost exhaustive combination of all possibilities of this system by the *unconscious ear* of the creative musician, by his intuition. Still absolutely missing is the correct identification of the relation between what has been attained and what is still to strive for. We must yet strive for everything that is left over.[25]

The twelve-tone method that operates on the basis of a unique, always new arrangement of tones for each musical composition is the ultimate, technical expression of such a striving. A large component of its logic is the act of *liquidation.* In a very similar manner to what is known in psychoanalysis as 'abreaction' (the process of 'wearing-away'), liquidation gradually eliminates certain acoustic characteristics, dissolving them, until what remains does not demand continuation. This is the point where its leftovers can be used in new configurations and the process can begin all over again.

> In dissolution the most important thing is to let go as quickly as possible of everything characteristic, to allow tensions to ebb, and so to neutralize the obligations of the earlier gestalten [shapes] as to liquidate, so

> that a clean slate, so to speak, is effected, providing the possibility for something different to come forward.[26]

This is how new coordinates for what is left over are found, through interminable condensation and displacement of sonic materials. It is the force of movement and developing variation that is the indispensable, creative moment here. Not unlike the psychoanalytic session, variation is maintained until the internal exhaustion of its momentum is reached. And it is only the ear that can take the leadership role in the intuitive imagination: "My ear has said: yes, and the ear is, after all, a musician's whole understanding."[27]

This understanding is not concerned with knowledge but perception. This is how Freud and Schoenberg construct a psychology and aesthetics of 'wear and tear.' Freud's tear initially assumes the shape of the famous *navel* of a dream. The navel is a point of contact, like a nerve ending, the "spot where it reaches down into the unknown" (SE 5: 525). It is a soft spot, a never healed wound, a channel of communication that emerges during analysis, indicating where the search should begin. And this is where interpretation can also end, either because there is not enough signifying material at hand, or because one plummets into the depths too recklessly.

The building blocks of dreams and tones are the openings and breaches in the psychological and aesthetic fabric. Schoenberg uses practically the same metaphor:

> I have shown that the system of tertian harmony has a rupture and that chords that do not fit in the system have been grouped together under the heading 'non-harmonic tones.' I have exposed this classification as the poorly disguised attempt to plug up the hole in the system with an imposing heap of unexamined scrap material, a heap so big that neither the hole nor the system itself is big enough to accommodate it.[28]

It is thanks to the peculiar movement through these loops and breaches that the decision of where to cease is so difficult: "Thus, we may well ask: Why, in what manner, and when does a piece of music close? The answer can only be a general one: As soon as the goal is reached."[29] But there is a difference between arriving at a point of conclusion and a mere arresting of motion: "Stopping is simple; it means not continuing. Closing, though, is different. To close, one must use special means."[30] Such means is found in the conscious ego as the *echo* of all these vibrating and resonating components. The same is true of language, because "words, since they are the nodal points of numerous ideas, may be regarded as predestined to ambiguity" (SE 5: 340). This is why "the spelling of words [or their visible surface] is far less important than their sound" (SE 5: 406).

Words lose their weight as mere speech-phenomena and become acoustic configurations of meaning, expressing "the peculiar plasticity of the psychical material [which] must never be forgotten" (SE 5: 352). It is perhaps only now that we can understand Freud's personal relationship to the art of sound, or rather why it is that he cannot take much pleasure in it, as he says in *The Moses of Michelangelo*:

> Works of art do exercise a powerful effect on me, especially those of literature and sculpture, less often painting. This has occasioned me, when I have been contemplating such things, to spend a long time before them trying to apprehend them in my own way, i.e., to explain to myself what their effect is due to. Wherever I cannot do this, as for instance with music, I am almost incapable of obtaining any pleasure. (SE 13: 211)

This remark places Freud somewhere 'beyond the pleasure principle' in relation to sound, leaving much for speculation. The essay appeared anonymously in 1914, and remained so for ten years, when the disguise was finally lifted in 1924.[31] The key seems to be contemplation—a type of duration that gives pleasure where sound cannot, because contemplation literally gives more time. Sound requires a special means of attention, precisely the type of attention that lies at the heart of the psychoanalytic method, which does not discriminate and is 'evenly suspended'. And, in fact, as one reads the text of *Moses*, Freud indulges in an almost pedantic interpretation of every detail of the sculpture (the position of the hands, the finger in the beard, the way the stone tablets are wedged under the arm, etc.), inordinately so—an analysis that in the end goes against the grain of the psychoanalytic technique. Sound is too close to the neutral plastic element, the driveshaft of the psyche, making it a matter of professional, technical, and scientific consideration (even though all this always remains implicit in Freud) and not pure aesthetic enjoyment. This communicative prowess of sound, as Schoenberg writes in 1912, is "the most wonderful thing about music . . . that one can say everything in it, so that he who knows understands everything; and yet one hasn't given away one's secrets—the things one doesn't admit even to oneself."[32] That such secrets can be given away by a special technique of amplification and listening is the lesson depth-psychology teaches. And it is interesting that in 1933 when attempting to describe the operatic treatment his Moses receives in *Moses und Aron*, Schoenberg also evokes Michelangelo's sculpture to indicate something that lies beyond the merely human and its psychology. This is the same gesture of 'reaching-beyond' the horizon of events we have already encountered—something that pushes perception beyond the limits of understanding.[33]

Even if we do not find pleasure in music as such, there is definitely something to be won through the analysis of sound. To arrive at the forming element itself, the element that animates sound, is to remake the tone into a new way of thinking, just like "at bottom, dreams are nothing other than a particular form of thinking" (SE 5: 506). To speak in those terms is to once again acknowledge in agreement with Freud that dream-work "does not think, calculate or judge in any way at all; it restricts itself to giving things a new form" (SE 5: 507). Schoenberg echoes this kind of sentiment: "Invention, but not calculation! One may compose by taking thought, but one must not deliberately observe how one is thinking. One can create freely within a tonality (*Tonart*) only if there is the feeling for this tonality in the unconscious."[34]

Ultimately Schoenberg's sound and Freud's psychoanalysis are the two most powerful instances of the redesigned topography of perception and thought. Thinking through this type of space does not proceed from abstract, clearly focused centers. On the contrary, it is a space one must inhabit:

> We have already become acquainted with the interpretative rule according to which every element in a dream can, for purposes of interpretation, stand for its opposite just as easily as for itself. We can never tell beforehand whether it stands for the one or the other; *only the context can decide* . . . This turning of a thing into its opposite is made possible by the intimate associative chain which links the idea of a thing with its opposite in our thoughts. (SE 5: 471, my emphasis)

The importance of the context, the fact that it is both the impetus for complexes of tones and psychic events, establishes something along the lines of *Tonarbeit* (an extension of the logic of *Traumarbeit*), or tone-work—already evident in Schoenberg's own description of his method. It is not 'atonality' or later the twelve-tone technique. Schoenberg's much more cumbersome but accurate designation is a method of composing with tones related only with one another; a definition he insists on throughout.[35] This is how tones become occurrences, not mere pitch frequencies through sequencing and repetition, creating associations that in turn lead to new symbolism, a new identity in sound. This kind of coherence constantly diversifies its expression. Hence, it is not a simple convergence but a *creative mechanics* of aesthetic forms.

Tonality creates a false space, whose volume is quickly overfilled, because it makes censorial decisions based on the eye and its always limited capacity for simultaneous reception of information:

> I maintain, namely, that the system of triads was extended to chords of four and five tones only because and wherever the image of superimposed thirds allowed extension, addition of thirds on top of thirds. Actually, then, [this extension grew] out of the graphic representation of the sounds, out of the visual impression: point, space, point, space, point, space . . . Only the sounds that could be traced back to this image [notation] were incorporated into the system.[36]

Standard harmony, according to Schoenberg, is filled to the brim with such artificial solutions that result in arbitrary exclusions on the basis of visual representation. Schoenberg worked feverishly against such a visual solution to sound—to the extreme, as is highlighted by the following anecdote. Apparently, when in 1908 Schoenberg showed the score of his *Five Orchestral Pieces op. 16* to Mahler, the latter openly stated "that he could not read it, for he was unable to translate the optical impression of the pieces into the acoustical [!]"[37]

Tonality instills artificial geometries, organizing and polishing musical reality, like the visual perspective sets up stable intervals of distances, of enlargement and diminishment. But in the long run any space that frames content in such a seemingly rational manner distorts the entire environment of which it is the expression, precisely because it cannot express much beyond the pure application of its technique. Whether ultimately Schoenberg himself falls into such a mannerism of administered technique, as many commentators including Adorno thought, is a matter of debate. But the question is very pertinent, perhaps more than ever, given that we live in the age of absolute design, from the software that runs our information to organic modification in the biosciences. At that time, this question was being asked in art and architecture, and there might be some important lessons still to be learned from their analytical and historical exploration.

But the goal, as we might suspect by now, was precisely the opposite: to enliven space, fertilize it by organic-like configurations, whose living tissue can accept transplants and be used in a sort of organ-donation model that supports other realities, which need its vitality. It is such analytical-organic space that underlies the texture of Schoenberg's sound. It is made up of multiple dimensions, not only the pitch, melody, or the cadence-driven harmony but also other elements that interact simultaneously, on all planes and levels.[38] The architectonics of such a space demands a lot of perception, requiring a new mode of sensorial approach:

> *The unity of musical space demands an absolute and unitary perception.* In this space, as in Swedenborg's heaven (described in Balzac's *Seraphita*) there is no absolute down, no right or left, forward or backward. Every musical configuration, every movement of tones has to be comprehended primarily as a mutual relation of sounds, of oscillatory vibrations, appearing at different places and times.[39]

To work and perceive in this mode, one needs a new type of awareness and a new form of consciousness. In his 1912 manifesto, *On the Nature of Visions*, Oskar Kokoschka expressed a very similar constructive logic:

> This change in oneself, which follows on the vision's penetration of one's very soul, produces the state of awareness, of expectancy. At the same time there is an outpouring of feeling into the image which becomes, as it were, the soul's plastic embodiment. This state of alertness of the mind or consciousness has, then, a waiting, receptive quality . . . The

> life of consciousness is boundless . . . When we no longer inhabit our perceptions they do not go out of existence; they continue as though with a power of their own, awaiting the focus of another consciousness. There is no more room for death; for though the vision disintegrates and scatters, it does so only to reform in another mode.[40]

In other words, one operates here as though in a dream, if only because it is ultimately consciousness as the organ of perception that registers its multiple choreography: "the desire for a conscious control of the new means and forms will arise in every artist's mind; and he will wish to know *consciously* the laws and rules which govern the forms which he has conceived '*as in a dream*.'"[41] This is not empty desire, because it is matched in its parameters by the general model of the psyche: "dream-work is under some kind of necessity to combine all the sources which have acted as stimuli for the dream into a single unity of the dream itself" (SE 4: 179). This unity is achieved by virulent juxtapositions:

> Thus speeches in dreams have a structure similar to that of breccia, in which largish blocks of various kinds of stone are cemented together by a binding medium. (SE 5: 419)

The binding agent that connects this bric-a-brac is nothing else than the simultaneity of occurrences, the bare-bones minimum of any harmony (precisely the threshold from which Schoenberg starts out) inscribed in the method of association:

> Our perceptions are linked with one another in our memory—first and foremost according to simultaneity of occurrence. We speak of this fact as 'association'. (SE 5: 539)

In the course of discussing chorale harmonization, Schoenberg presents such a type of generic organicism as an example of how tones are combined and harmonies created, when things are taken down to the level of least resistance:

> Such mosaic-like assemblage of parts permits no very complicated relationships and favours as its unifying element, the principle of more or less simple repetitions . . . the unconnected parts have no particular duty toward one another; perhaps they are on the whole just next to one another more than in relation to one another.[42]

Schoenberg's theory and practice rest almost exclusively on this notion of juxtaposition and extension of boundaries. Harmony thus becomes a psychology of transgressed limits. It is a question of final horizons—their maintenance and, when needed, violation. This is the only way to transmit something that would otherwise remain beyond the grasp of perception—the ultimate justification for art, whatever the medium.[43] The desire for the elsewhere, the encounter beyond the vanishing point or the horizon line, is an archetypal experience that stands as an emblem of the entire 'out-of-bounds' complex in Freud, Schoenberg, and their cultural environment.

Conclusion
The Passion for the Real

The task before us has been not only to show an area of convergence between Freud and Schoenberg as two exemplary personalities of the *fin de siècle*, but through them to get a sense of the logic and the spirit of that time, of which we are the intellectual and cultural descendants. By necessity it was a limited encounter. But, even within these limits, we found surprising overlaps in thinking about identity and expression in fields that technically speaking should be on opposite ends of the perceptive spectrum, and have been, more often than not, treated this way in academic discussions. I have attempted to approach these as much as is possible through the mechanism of simultaneity and the 'all-at-once' presence of widely divergent elements—the same ones that were indispensable to Schoenberg's sound (more and more as he advanced in his craft) and the entire *fin de siècle*.

Simultaneity of occurrence and immediacy of understanding have again taken center stage, or maybe it would be more accurate to say that they have never bowed out. Just as Schoenberg searched for a unified theory of perception without which, he thought, no meaningful sound could take shape, so at present, when things are continually broken apart into ever smaller particles of space and matter, we seem to be in as urgent a need as ever of a new sensory arrangement—something that would also recalibrate perception, finding connections between the various dimensions of presence and the increasingly elastic distances between them.

If nothing else, Schoenberg's uniquely designed aesthetic awareness was precisely this type of distance mechanics that recalibrated perception on the largest scale possible:

> There is a certain distance, from which we can really detect only the spirit (*Geist*) of the century. He who can set himself far enough apart will detect the spirit of mankind. Individualities disappear at this distance, but what they express—mankind, the best that is in it—becomes visible. The highest pinnacles, which are most accessible to the observer, into which the capillaries lift the finest and best from the depths, these alone set forth the spirit of mankind. Thus, increasing distance, initially reductive, once again magnifies: the individuals, the pinnacles,

> become visible again, even if in a different way. One sees that they are related and how they are related, that they are coherent among themselves; one no longer sees what proximity revealed, one no longer sees that they were sharply separate; *but the relationships are not those of art, not those of the techniques of art; they are rather deeper ones.*[1]

The search for deeper relationships, the *modus operandi* at the time, involved a type of sensitivity that perceives anew, because it finds a different range of empiricism that folds events and spaces, including time, one into the other: "Perhaps the future of our music speaks through this freshness [of new chords]. It is heard only by those who are highly sensitive to impressions, the impressionists. The impressionist's organ [of perception] is a mechanism of extraordinary fine tuning, a seismograph, which registers the slightest motion."[2] Such new-age impressionism takes as its building blocks the many breakdowns in instrumental rationality—failed connections (*Fehlleistungen*) that express the deeper and invariant elements of phenomena. Only this operation rouses towards a reconstruction of reality, according to Freud:

> For we must always be prepared to drop our conceptual scaffolding if we feel that we are in a position to replace it by something that approximates more closely to the unknown reality. (SE 5: 610)

Schoenberg is the model of such a demolition where conceptual scaffolding (in his case tonality) is dropped and remodeled, initially through the so-called, impulse-driven atonal stage and later increasingly compressed into the mobile but more solidified schematics of sound as sequences of tones.

Parapraxis, the cumbersome and inscrutable translation of *Fehlleistung* commonly known as the 'Freudian slip', was introduced by Freud in 1901, in one of his most widely read books—*The Psychopathology of Everyday Life*. In the course of his exposition, Freud shows failed connections as deregulated distance management procedures that expose the senses to the proximity of something previously far removed. Slips of meaning and form bring closer the far-off element that is otherwise inapprehensible, while still keeping it far enough to fend off the potentially dangerous consequences of overmagnification and exaggerated proximity. In other words, they set the entire distance-proximity matrix into motion. Schoenberg's entire musical output is a conglomeration of such failed connections, not only because these are the connections that would not be allowed under the standard system of harmony but also because they are driven by a deeper logic of misrecognized impulses. Contrary to all appearances, tones in Schoenberg's system are not free-flowing entities (just like the psyche is not) but connected and bound by a space that in itself is a surface on which countless displacements are allowed to take place. Standard English translation, again very clumsily, refers to this type of movement and its dynamics in psychoanalysis as cathexis, whereas Freud simply calls it *occupation* (*Besetzung*).

This is the case, for instance, with the seventh chord: "The seventh is an aggressive tone [*ein Strebeton*], that is, a tone exhibiting the drive [*ein Ton, der zwingenden Trieb zeigt*] toward resolution into another.[3] All occupation is aggressive, motivated by the desire to be otherwise, or somewhere else, and to possess these conditions. Hence the *Strebeton* is in fact closer to the 'desiring tone' that is under the compulsive drive (*zwingender Trieb*) towards a resolution in another. Schoenberg sees all tones as such aggressive/desiring 'machines' that always aspire towards an occupation of a different space, because they consistently lead away from the tonal center, no matter how hard one attempts to control them. Occupation, then, is an economic system—a relationship that displaces energies form the mute unconscious, and amplifies them in an intense and condensed manner by making them pass through other configurations, phenomena, and objects. This is why occupation is the energetic principle behind all symptom-building—something like 'the passion for the real' or that desire and hubris so characteristic of the last century, of wanting "to make man here and now."[4]

***

Just a few decades after the ignition point for what has been called 'hypermodernity' in the years around 1900, the voice takes over almost completely. It is found in disembodied networks of communication, in politics as propagandistic aesthetics and ideology, in the postwar thought as pure discourse, in poetry and literature. "Perse made the 'I' and the 'we' equivalent, but in Celan's poem [*Anabasis*], as in Beckett's prose, there is no longer either 'I' or a 'we'. There is just a voice trying to trace a way."[5] This voice loses its prior sonic quality. It becomes hollow, an isolated tone that is put in charge of redeeming the emptiness around it, instead of *using it*, as one still did at the turn of the century, to reshape the contents of reality. In other words, one lost the willingness and the ability to inhabit the ruptures and breaches in meaning.

Art today seems to be grotesquely banal in comparison with the charge it arrogated to itself (for better or worse) at the turn of the twentieth century, where it became the medium disseminating the logic of decay. This is certainly the conclusion of Paul Virilio, who speaks of the 'banalization of excess' and through it the loss of the sense of value:

> Without limits, there is no value; without value, there is no esteem, no respect and especially no pity . . . Already, more or less everywhere you turn, you hear the words that precede that fatal habituation to the banalization of excess. For certain philosophers the body is already no more than a phenomenon of memory, *the remnants of an archaic body*.[6]

The *fin de siècle* prefigured and maybe even liberated the psychopathology of violence the twentieth century carried through in the brilliant glow of scientific and technological prowess. Aesthetics, as a complete system of perception and its cultural legitimacy as art, seems to have lent a helping hand. We need only think here of the futurists, who exalted the beauty of war, or the decadent 'poetry' (if indeed decadence can still be poetic) of someone like the Czech poet Jiří Karásek ze Lvovic, who wrote in a dark room, surrounded by the stench of decaying flowers. But this story, even if it makes for spectacular literature or scholarship, does not end, just as it did not begin there. This is why I disagree with Virilio's polemics of pessimism that sees only presence and thus no possibility—a pitiless present of inhuman and dumbfounding immediacy that cuts into experience like a blunt knife. As we now realize, there is another side to the story, the one I have attempted to delineate—the story of a project for a new psychology and through it, a new mode of perception that would give modern men and women a means of dealing with the always inescapable and necessary removal from the reality that renders us incapacitated.

This is what the series of conceptual formulations (my own narrative's complex) attempted to untangle, as they complicated—a series of themes that in the manner of Schoenberg's thematic variation was repeated in mirroring shapes, echoes, and inversions. Each one of these formulations tried to carry the subject matter within its own confines, in this sense mimicking what Anton von Webern said about Schoenberg's D minor quartet: "There is no note, so to speak, in this work which is not thematic."[7]

In this manner, we are again at the head of what faces us continually, the awareness of our own situation from which we must be able to recover, because "if we could look at the present from afar, then all our struggles would be over."[8] It is precisely this type of distance, this kind of procedure that inserts an interval between the present and its immediacy Freud and Schoenberg attempted to recalibrate, and whose failure Paul Virilio bemoans as the new tyranny of the pitiless (art that has lost its power of representation), the inhuman, the nonemphatic (aesthetics that completely eliminates ethics). But all this says is that aesthetics and art as such are not the spaces where the search can be carried on. It does not eliminate the struggle between opposing elements that must find a way of interacting beyond the prosaic forms of communication. And, in turn, it indicates that today more than ever we are still lacking a *psychology of distances* that would allow for perception that is not simply a mute registry of presence but rather a space of active encounters, where one feels some responsibility for culture's overall design.

Notes

NOTES TO THE INTRODUCTION

1. Kundera, *The Art of the Novel*, 124.
2. See Blom , *The Vertigo Years: Europe 1900–1914.*
3. The complicated and fascinating history of the predominance of vision in Western imagination and its later erosion as the predominant sense in artistic and intellectual milieu of French thought is told by Martin Jay in *Downcast Eyes: The Denigration of Vision in Twentieth-Century French Thought.*
4. See Erlmann, *Reason and Resonance: A History of Modern Aurality.*
5. One way was through the movement and general perceptive disorientation of decadence that in the end paradoxically also proved to be a generative-productive condition, its ambiguities notwithstanding. For a discussion of decadence in music see Downes, *Music and Decadence in European Modernism: The Case of Central and Eastern Europe.*
6. Erlmann, *Reason and Resonance*, 132.
7. Friedrich Hebbel, quoted in Schorske, *Fin-de-Siècle Vienna: Politics and Culture*, xxvii.
8. This is the central thesis of the now classic exploration of Viennese culture by Schorske.
9. See Schorske, *Fin-de-Siècle Vienna*, xvii–xviii.
10. Mach, *The Analysis of Sensations, and the Relation of the Physical to the Psychical*, 4.
11. Once again, the seminal study that initiates a long chain of such engagements in the English-speaking scholarship is Schorske's *Fin-de-Siècle Vienna.* Schorske essentially rediscovers Vienna's unique character as the crucible of modern times.
12. As far as I am aware, Prague is the only city where Cubism found materialization not only on canvas, but through architecture. It also holds a very special place in the Surrealist movement, which was practically absent from Vienna. On this topic and the general importance of Prague to European artistic culture (a history that disappeared from West's consciousness for fifty years after 1945) see Sayer, *The Coasts of Bohemia: A Czech History*, and *Prague, Capital of the Twentieth Century: A Surrealist History.*
13. Freud, 'Screen Memories', in *The Standard Edition*, 3: 318. Hereafter cited in text as SE.
14. Arnold Schoenberg to Alma Mahler, 27 March 1911, in *A Schoenberg Reader: Documents of a Life*, Joseph Auner ed., 96. Freud, not unlike Schoenberg, was also very disparaging in his comments about the city throughout his life. See Gay, *Freud: A Life for Our Time*, 9–10, and Jones, *Sigmund Freud: Life and Work*, 1: 322.

15. See Musil, *The Man without Qualities*. As is often the case, the English translation of *Der Mann Ohne Eigenschaften* is misleading. The German indicates not so much a lack of qualities, but the absence of character traits that are uniquely one's own. It suggests a dispossession or even an absence of personal characteristics as such. Musil builds his entire narrative around this principle of dissipation, where the protagonist, Ulrich, is an effigy of an individual—his psychological elasticity bringing him closer to the blurred edge, spanning mere matter, and an unidentifiable human element.
16. See Badiou, *The Century*.
17. Kraus, *Dicta and Contradicta*, 105.
18. Schorske, *Fin-de-Siècle Vienna*, 340.
19. See the introduction to Schoenberg's *Coherence, Counterpoint, Instrumentation, Instruction in Form*, lvii–lix.
20. *Ibid.*, 61.

NOTES TO CHAPTER 1

1. Agamben, *The Time That Remains*, 131.
2. Gay, *Freud: A Life for Our Time*, 13–14.
3. Stuckenschmidt, *Schoenberg, His Life, World and Work*, 20.
4. In fact, between the years of 1885 and 1886, Freud published three neurological papers, whose central concern was the acoustic nerve. A good synopsis is found in Jones, *Sigmund Freud, Life and Work*, 1: 225–227.
5. In the initially anonymous paper, published in 1914, *The Moses of Michelangelo*, Freud writes, "Whenever I cannot do this [contemplate a work of art], as for instance with music, I am incapable of obtaining any pleasure (SE 13: 211). I will discuss this paper, and the unusual circumstances surrounding it, at a later point.
6. Hanslick, *The Beautiful in Music*, 78–79, hereafter cited in text.
7. Hermann von Helmholtz, quoted in Warren and Warren, eds., *Helmholtz on Perception*, 83.
8. Interestingly, in a reverse motion, the aural was applied to the descriptions of the new phenomenon of photography: "The Photographer needs in many cases no aid from any language of his own, but prefers rather to listen, with the picture before him, to the silent but telling language of Nature" (1833). William Henry Fox Talbot, quoted in Georges Didi-Huberman, *Invention of Hysteria*, 32.
9. Jankélévitch, *Music and the Ineffable*, 11, hereafter cited in text.
10. On the theme of autobiography, autoanalysis, psychoanalysis, and music, especially through the writings of Theodor Reik (a member of the Freud circle), see Lacoue-Labarthe, "The Echo of the Subject".
11. Nancy, *Listening*, 13.
12. Anna O's story is missing from the first two German editions of Freud's complete works! See editor's introduction to *Studies on Hysteria* (SE 2: x).
13. Nancy, *Listening*, 20.
14. For an excellent study of hysteria's genealogy, especially its nineteenth century reinvention in the Charcot clinic, see Georges Didi-Huberman, *Invention of Hysteria: Charcot and the Photographic Iconography of the Salpêtrière*.
15. Schoenberg, quoted in Auner, ed., *A Schoenberg Reader*, 118.
16. This idea was well established in the medical circles at the time. In fact, the self-professed great inventor of hysteria in the nineteenth century, Jean-Martin Charcot, devoted his thesis for the professorship exam (*aggrégation*) to the concept of expectation. See Didi-Huberman, *Invention of Hysteria*, 102.

NOTES TO CHAPTER 2

1. Kittler, *Discourse Networks 1800/1900*, 13.
2. Pushkin, quoted in Lukács, *Goethe and His Age*, 157.
3. In a remarkable passage, Freud indicates that he will not accept a patient towards whom he does not feel sympathetic: "I cannot imagine bringing myself to delve into the psychical mechanism of a hysteria in anyone who struck me as low-minded and repellent, and who, on closer acquaintance, would not be capable of arousing human sympathy" (SE 2: 265).
4. As it turns out, Strindberg's work constitutes one of the most important creative influences on Schoenberg. In a 1909 letter to Alban Berg, Schoenberg writes, "Just see that you don't take the *Dream Plays* away from me, for I'm considering them myself. But some other Strindberg work! I consider that very feasible!" (quoted in Joseph Auner ed., *A Schoenberg Reader*, 117).
5. Schoenberg, "New Music: My Music", in *Style and Idea*, 105.
6. There is conflicting evidence as far as any family ties between the two women are concerned.
7. Wellesz, *Arnold Schoenberg: The Formative Years*, xii.
8. All of Schoenberg's paintings come from the period between 1907 and 1910, culminating in an exhibition of his works in the autumn of 1910 at the Heller's Gallery in Vienna. Interestingly, Schoenberg also painted the portrait of his young librettist, Marie Pappenheim.
9. Wellesz, *Arnold Schoenberg: The Formative Years*, ix.
10. Schoenberg, *The Musical Idea*, 119.
11. *Ibid.*, 153.
12. See Dahlhaus, "Schoenberg's Poetics of Music", in *Schoenberg and the New Music*, 73–80.
13. See Adorno, *The Philosophy of New Music.*
14. Schoenberg, *The Musical Idea*, 153.
15. *Ibid.*, 107.
16. Freud, "Some General Remarks on Hysterical Attacks", SE 9: 229.
17. Adorno, *The Philosophy of New Music*, 40.
18. See Simms, *The Atonal Music of Arnold Schoenberg 1908–1923*, 97–99.
19. Dahlhaus, *Schoenberg and New Music*, 153–154.
20. Schoenberg, *The Musical Idea*, 121–123.
21. See Rognoni, *The Second Vienna School, Expressionism and Dodecaphony*, 1–2. Such intellectual orientation, like many conceptual complexes at the time, goes back to Goethe's idea of the '*ur*-phenomenon'.
22. See Breton's *Nadja.*
23. Nietzsche, *On the Uses and Disadvantages of History for Life*, 62.
24. Freud, *Analysis Terminable and Interminable*, SE 23: 241.
25. Nietzsche, *Zarathustra*, 13.
26. For an excellent anthology of some of the most significant manifestos written around that time, and well into the last century, see Mary Ann Caws, ed., *Manifesto: A Century of''Isms'.*
27. This line of progression is more or less contained in Michel Foucault, *Discipline and Punish, Madness and Civilization* and *The History of Sexuality.*
28. Wilhelm Reich, quoted in Virilio, *Aesthetics of Disappearance*, 43.
29. Kandinsky, *Point and Line to Plane*, 31.
30. See Düchting, *Paul Klee: Malerei und Musik.*
31. Van Gogh's paintings were exhibited for the first time in Vienna, in 1903 at the Secession's Exhibit of Impressionism. On Van Gogh's impact on Viennese modernism, see Krapf-Weiler's essay "The Response of Early Viennese

Expressionism to Vince van Gogh", in Werkner, ed., *Egon Schiele: Art, Sexuality, and Viennese Modernism*, 31–50.
32. Nietzsche, *Human All Too Human*, 13–14.
33. Bahr, *Expressionismus*, 20–21, my translation.
34. See Erlmann, *Reason and Resonance*, 31 .
35. See *ibid.*, 31 for a fascinating discussion of *inconcussum*, usually translated as 'indubitable' but immersed in a rich etymological field from *cutere* (to shake violently), creating such amalgams as percussion, discussion, concussion, etc.
36. For a methodical explication of Descartes's thought as an elaborate system of linear, discursive, and aesthetic extension, see Claudia Brodsky Lacour's *Lines of Thought: Discourse, Architectonics and the Origin of Modern Philosophy.*
37. Descartes, *Meditations*, 71.
38. Brodsky Lacour, *Lines of Thought*, 149.
39. *Ibid.*, 37.

NOTES TO CHAPTER 3

1. Bahr, *Expressionismus*, 24, hereafter cited in text. All translations are mine.
2. See Rognoni, *The Second Vienna School: Expressionism and Dodecaphony*, who traces an interesting genealogy of the term, attributing it at least in part to Wilhelm Worringer's essay on Cézanne, Van Gogh, and Matisse published in 1911 in the literary review *Der Sturm*. Worringer seems to have been the first to use 'expressionism' as a critical term. However, some historians also attribute the term to Julien-Auguste Hervé, who used it in reference to a group of his own paintings on view at the 1907 'Salon des Indépends' in Paris. On the general concept and the movement see also Crawford and Crawford, *Expressionism in Twentieth-Century Music*.
3. Stuckenschmidt, *Schoenberg*, 98, 143.
4. Wistrich, *Laboratory for World Destruction*, 310.
5. Badiou, *The Century*, 37.
6. For a very interesting and compelling discussion of how modernity in the nineteenth century begins to define experience in terms of how one pays attention, and the vicissitudes as well as corruptions associated with it, see Crary, *Suspensions of Perception: Attention, Spectacle, and Modern Culture.*
7. See Uexküll's *Umwelt und Innenwelt der Tiere.*
8. Uexküll, "An Introduction to Umwelt", 107.
9. See Tarasti, "Metaphors and Nature and Organicism in the Epistemology of Music", 664.
10. Mach, "Antimetaphysische Vorbemerkungen", quoted in *Die Wiener Moderne*, 141, my translation.
11. *Ibid.*, 138.
12. Rilke, *Letters to A Young Poet*, 20.
13. Panofsky, *Perspective as Symbolic Form*, 40.
14. Derrida, "Force and Signification", in *Writing and Difference*, 279.
15. Damisch, *The Origin of Perspective*, 39.
16. Merleau-Ponty, *Visible and the Invisible*, 201.
17. See Rosen, "The Ruins of Walter Benjamin", in Gary Smith, ed., *On Walter Benjamin*, 134.
18. See Wood, *Vienna School Reader: Politics and Art Historical Method in the 1930s*, 14.

19. Reigl, "The Main Characteristics of the Late Roman *Kunstwollen*", in Wood, ed., *Vienna School Reader*, 93.
20. Panofsky, *Perspective as Symbolic Form*, 41–42.
21. Goethe, quoted in Nietzsche, *On the Uses and Disadvantages of History for Life*, 59.
22. Panofsky, *Perspective as Symbolic Form*, 71–72.
23. Schorske, *Fin-de-Siècle Vienna*, 346.
24. Schoenberg, *The Musical Idea*, 119.
25. Riegl, *Historical Grammar*, 137–138.
26. *Ibid.*, 139.
27. See Jacqueline E. Jung, preface to Riegl, *Historical Grammar of the Visual Arts.*
28. Mach, *The Analysis of Sensations, and the Relation of the Physical to the Psychical*, 4. Mach's book, first published in 1886, went through five German editions, the last one in 1906.
29. Riegl, *Historical Grammar*, 161.
30. Riegl, *Historical Grammar*, 361–362.
31. Bahr, *Expressionismus*, 54.
32. See Stuckenschmidt, *Schoenberg, His Life, World and Work*, 93.
33. See my "Bending Modernity: Psychoanalysis and the Rest of Culture."
34. Blanchot, The Writing of the Disaster, 51.

NOTES TO CHAPTER 4

1. Mersenne, quoted in Erlmann, *Reason and Resonance*, 104.
2. Riegl, *Historical Grammar*, 294.
3. *Ibid.*, 300.
4. Badiou, *The Century*, 37.
5. Badiou, *The Century*, 45.
6. Lecture XXXV, "The Question of a Weltanschauung", in *Introductory Lectures to Psychoanalysis*, SE 22, hereafter cited in text.
7. Wittgenstein, *Wittgenstein's Lectures, Cambridge 1932–1935*, 39–40. On the general theme of Wittgenstein's philosophy and psychoanalysis see Bouveresse, *Wittgenstein Reads Freud: The Myth of the Unconscious.*
8. Nancy, *Listening*, 11.
9. See Schoenberg, *Theory of Harmony.*
10. See Christensen, *Rameau and Musical Thought in the Enlightenment*, 107–109.
11. Nancy, *Listening*, 21.
12. See Virilio, *Aesthetics of Disappearance.*
13. Schoenberg, *The Musical Idea*, 115–117.
14. This is also the modern German noun for an actor.
15. This is the part of Freud that Jacques Lacan recuperates and reinvents in his project of going back to the roots of Freudian theory, under the concept of *jouissance* (of play, game, enjoyment, etc.).
16. The term *das Es*, as Freud himself explains (SE 19: 23) goes back in the first place to Georg Groddeck, and then certainly back to Nietzsche. See also "Editor's Introduction" (SE 19: 7).
17. William Burroughs, quoted in Virilio, *Aesthetics of Disappearance*, 67.
18. Alois Riegl, *Historical Grammar of the Visual Arts*, 137.
19. The entire episode is described in *Beyond the Pleasure Principle*, SE 18: 14–17.
20. See Szendy, *Listen: A History of Our Ears.*

NOTES TO CHAPTER 5

1. Serres, *The Parasite*, 121.
2. Adorno, "On the Fetish-Character in Music and the Regression of Listening", in *Essays on Music*, 289.
3. Schoenberg, *Style and Idea*, 244.
4. Nancy, *Listening*, 4.
5. *Ibid.*
6. Adorno, "On the Fetish Character in Music", 291.
7. Schoenberg, *Style and Idea*, 215.
8. Nancy, *Listening*, 9.
9. *Ibid.*, my emphasis.
10. Marshall McLuhan, quoted in Cavell, *McLuhan in Space*, 91.
11. Strindberg, *Selected Essays*, 180.
12. For an excellent discussion of Strindberg's artistic activity, intermixed with biography and general cultural commentary on the times, see Robinson's "Introduction", in Strindberg, *Selected Essays by Strindberg*, 1–21.
13. See Krapf-Weiler, "The Response of Early Viennese Expressionism to Vincent van Gogh", in Werkner, ed., *Egon Schiele*, 31–50.
14. Strindberg, *Selected Essays*, 180.
15. *Ibid.*
16. See Stuckenschmidt, *Schoenberg, His Life, World and Work*, 239.
17. *Ibid.*, 233.
18. See *ibid.*, 233–239.
19. Strindberg, *Selected Essays*, 250.
20. *Ibid.*
21. Strindberg, "Deranged Sensations", in *Selected Essays*, 122–123.
22. Artaud, quoted in Caws, ed., *Manifesto: A Century of 'Isms'*, 459.
23. Strindberg, "Deranged Sensations", in *Selected Essays*, 123.
24. *Ibid.*, 124.
25. Focillon, *The Life of Forms*, 79.
26. Strindberg, "Deranged Sensations", in *Selected Essays*, 125.
27. *Ibid.*, 127.
28. *Ibid.*, 128.
29. See Stuckenschmidt, *Schoenberg*, 91.
30. Strindberg, "Deranged Sensations", in *Selected Essays*, 133.
31. *Ibid.*, 103–105.
32. *Ibid.*, 103.
33. *Ibid.*
34. *Ibid.*, 104.
35. Schoenberg, *Theory of Harmony*, 1.
36. *Ibid.*, 2.
37. *Ibid.*
38. *Ibid.*
39. *Ibid.*
40. *Ibid.*
41. The discourse of voicing and prying the mouth open was also extended to Freud, who was consistently referred to by the Viennese press as 'Sieg-mund' Freud (the victorious-mouth). For many instances of this, see Tichy and Zwettler-Otte, *Freud in der Presse: Rezeption Sigmund Freuds und der Psychoanalyse in Österreich 1895–1938*.
42. Sternberger, "Jugendstil—Begriff und Physiognomik", in *Jugendstil*, 28–29. The essay first appeared in 1934 in *Die Neue Rundschau* 2, 255–271. All translations mine.

43. *Ibid.*, 30.
44. *Ibid.*
45. *Ibid.*, 31.
46. *Ibid.*
47. *Ibid.*, 32, my emphasis.
48. *Ibid.*, 37.
49. *Ibid.*, 36, my emphasis.
50. *Ibid.*, 42–43.
51. Schoenberg, *The Musical Idea*, 109.
52. For a good discussion of organic sound and Western music, see Trasti's "Metaphors of Nature and Organicism in the Epistemology of Music", in *Sign Systems Studies*, 657–682.
53. See Frisch, "Music and Jugendstil", *Critical Inquiry* 17 (1990): 138–161.
54. Adorno, quoted in Frisch, "Music and Jugendstil", 160.
55. Sternberger, "Jugendstil—Begriff und Physiognomik", in *Jugendstil*, 45.
56. *Ibid.*, 46.

NOTES TO CHAPTER 6

1. Lukács, "Longing and Form", in *Soul and Form*, 111–127.
2. Alois Riegl engaged Semper throughout the *Historical Grammar of the Visual Arts*, often paying homage to one of the greatest theorists and architects of the nineteenth century, but also consistently moving beyond Semper's definition of art's origin in technique and the functionality of practical circumstances. Gottfried Semper (1803–1879) is known not only for his major and very influential work, *Der Stil* (Style), but also for his buildings. In the context of Vienna, the most visible marks left by Semper on its cityscape are the two mirror Neorenaissance buildings, facing each other across the Maria Theresa Square—*Kunsthistorisches Museum* and *Naturhistorisches Museum*, as well as the *Burgtheater*, facing the Town Hall across the Ring.
3. Semper, quoted in Moravánszky, "Aesthetics of the Mask", 205.
4. *Ibid.*, 208.
5. See Simms, *The Atonal Music of Arnold Schoenberg*, 114–119.
6. Moravánszky, "The Aesthetics of the Mask", 207.
7. *Ibid.*
8. Wagner, *Modern Architecture.*
9. *Ibid.*, 63.
10. Neumeyer, "Iron and Stone: 'The Architecture of the *Großstadt*", in Mallgrave, ed., *Otto Wagner*, 123, my emphasis.
11. Simmel, quoted in Frisby, *Fragments of Modernity*, 73.
12. Simmel, "Sociological Aesthetics", in Etzkorn, ed., *Georg Simmel—The Conflict in Modern Culture and Other Essays*, 71–72.
13. *Ibid.*, 68–69.
14. Simmel, "The Conflict in Modern Culture", in *The Conflict in Modern Culture and Other Essays*, 12.
15. Simmel, "Sociology of the Senses", in *Simmel on Culture*, 116.
16. Simmel, "Sociological Aesthetics", in *The Conflict in Modern Culture*, 70.
17. See *ibid.*, 77.
18. See my "Dwelling Within: The Inhabited Ruins of History."
19. Simmel, "Conflict in Modern Culture", 77–78.
20. Simmel, "On Aesthetic Quantities", in *The Conflict in Modern Culture*, 83.
21. Benjamin, "The Work of Art in the Age of Mechanical Reproduction", in *Illuminations*, 217–251.

22. *Ibid.*, 221, my emphasis.
23. *Ibid.*, 223.
24. *Ibid.*, 222–223, my emphasis.
25. *Ibid.*, 229, my emphasis.
26. *Ibid.*, 237.
27. *Ibid.*, 235, my emphasis.
28. Marinetti, quoted in Caws, ed., *Manifesto*, 187, 189.
29. Simmel, "The Ruin", in Kurt H. Wolff, ed., *Essays on Sociology*, 259.
30. *Ibid.*, 260.
31. *Ibid.*, 263.
32. *Ibid.*, 265.
33. *Ibid.*, 265.
34. *Ibid.*, 266.
35. Simmel, "The Essence of Culture", in *Simmel on Culture*, 41.
36. Simmel, "The Ruin", 262.
37. Benjamin, "The Work of Art in the Age of Mechanical Reproduction", in *Illuminations*, 220.
38. For a discussion and analysis of Wagner's early designs according to the paradigm of a 'ruin-in-reverse', see Berry, "From Historicism to Architectural Realism: On Some of Wagner's Sources", in Mallgrave, ed., *Otto Wagner*, 243–278.
39. Simmel, "Das Problem des Stiles", 307, my translation.
40. Benjamin, "The Work of Art in the Age of Mechanical Reproduction", in *Illuminations*, 240.
41. *Ibid.*
42. Neymeyer, "Iron and Stone", in Mallgrave, ed., *Otto Wager*, 129.
43. Bahr, *Kulturprofil der Jahrhundertwende*, 283, my translation.
44. Berry, "From Historicism to Architectural Realism", in Mallgrave, ed., *Otto Wagner*, 245.
45. Worringer, *Abstraction and Empathy*, 133–134.
46. *Ibid.*, 23.
47. *Ibid.*, 126.
48. Wagner, *Modern Architecture*, 121.
49. *Ibid.*, 62.
50. *Ibid.*,136n.
51. Jacques Rancière, *The Politics of Aesthetics*, 15.
52. Schoenberg, quoted in Stuckenschmidt, *Schoenberg, His Life, Work and World*, 157.
53. Loos, "Zwei Aufsätze und eine Zuschrift über das Haus auf dem Michaelerplatz", in Adolf Opel, ed., *Trotzdem: Gesammelte Schriften 1900–1930*, 113, my translation.
54. Loos, "Das Mysterium der Akustic", in *Trotzdem: Gesammelte Schriften 1900—1930*, 116, my translation.
55. *Ibid.*, 117, my translation.
56. Loos, "Von der Sparsamkeit", in *Ornament und Verbrechen*, 225–226, my translation.
57. *Ibid.*, 221, my translation.
58. Colomina, *Privacy and Publicity: Modern Architecture as Mass Media*, 65.
59. In the 1890s Loos literally went to the frontier, by spending several years in America, one of the first European modernists to do so. His sojourn there left an indelible mark on his thinking about space.
60. Loos, "Heimatkunst", in *Trotzdem*, 129, my translation.
61. Nancy, *Listening*, 6.
62. Loos, "Architektur", in *Trotzdem*, 96, my translation.

63. Colomina, *Privacy and Publicity*, 244.
64. Frisch, "Music and Jugendstil", 139.
65. Endell was the architect of the Jugendstil Bunte Theater in Berlin, where Schoenberg found his first employment in 1901.
66. See Helmholtz, *On the Sensations of Tone*.
67. See Lipps, *Consonance and Dissonance in Music*.
68. See Szendy, *Listen: A History of Our Ears*.
69. Carrà, quoted in Caws, ed., *Manifesto: A Centur of 'Isms'*, 204.
70. Russolo, quoted in *ibid.*, 207.
71. Tzara, quoted in *ibid.*, 301.
72. Caws ed., *Manifesto: A Centur of 'Isms'*, 289.
73. Tzara, quoted in *ibid.*, 288.
74. Aragon, Louis and André Breton, "Le Cinquntenaire de l'Hysterie (1878–1928)", *La Révolution Surrealiste*, 11 (1928), 20.
75. Endell, "Möglichkeiten und Ziele einer neuen Architektur", *Deutsche Kunst und Dekoration* 1 (1898): 141–152, my translation and emphasis.
76. Lipps, "Aesthetische Einfühlung", 444–445, my translation.
77. Lipps, "Aesthetische Einfühlung", 446–447, my translation.
78. Focillon, *The Life of Forms in Art*, 32.
79. Schoenberg, quoted in Stuckenschmidt, *Schoenberg, His Life, Work and World* 155.
80. Schoenberg, quoted in Crawford, "Die Glückliche Hand: Further Notes", 73.
81. Schoenberg, quoted in Stuckenschmidt, *Schoenberg, His Life, Work and World* 315, my emphasis.
82. Aars, "Die Erwartung", 401, my translation.
83. *Ibid.*, 403, my translation and emphasis.

NOTES TO CHAPTER 7

1. Zola, *The Experimental Novel and Other Essays*, 1.
2. Claude Bernard, quoted in Didi-Huberman, *Invention of Hysteria*, 20.
3. Zola, *The Experimental Novel*, 1–2.
4. *Ibid.*, 2.
5. *Ibid.*, 4.
6. *Ibid.*, 17.
7. *Ibid.*, 18.
8. *Ibid.*, 32.
9. See Schoenberg, *The Musical Idea*, 127.
10. Zola, *The Experimental Novel*, 23, my emphasis.
11. Stuckenschmid, *Schoenberg, His Life, Work and World*, 230.
12. Schoenberg, quoted in *ibid.*
13. Zola, *The Experimental Novel*, 38.
14. *Ibid.*, 54.
15. Musil, *Diaries*, 2–3.
16. Deleuze and Guattari, *What Is Philosophy?*, 60.
17. See Nietzsche, *The Birth of Tragedy and the Case of Wagner*, 19.
18. *Ibid.*, 34–35.
19. *Ibid.*, 34.
20. *Ibid.*, 64.
21. *Ibid.*, 52, my emphasis.
22. *Ibid.*
23. *Ibid.*

24. *Ibid.*, 61.
25. *Ibid.*, 40.
26. *Ibid.*, 60.
27. During the course of his discussion, Nietzsche cites an extensive fragment about music from Schopenhauer's *World as Will and Idea*. See *ibid.*, 101–103.
28. Nietzsche, *The Birth of Tragedy*, 104.
29. See Noudelmann, *The Philosopher's Touch*, 49–93.
30. Nietzsche, *The Birth of Tragedy*, 104.
31. *Ibid.*, 141.
32. *Ibid.*
33. Hofmannsthal, *The Lord Chandos and Other Writings*, 129.
34. Huter, "Body as Metaphor", in Patrick Werkner, ed., *Egon Schiele*, 124.
35. Hofmannsthal, quoted in Bennett, *Hugo von Hofmannsthal*, 109.
36. Hofmannsthal, *The Lord Chandos and Other Writings*, 123, translation modified.
37. *Ibid.*, 125, translation modified.
38. *Ibid.*, 120, translation modified.
39. Hofmannsthal, quoted in the introduction to Hugo von Hofmannsthal, *Poems and Plays*, xxiv.
40. Nietzsche, *The Birth of Tragedy*, 55–56.
41. *Ibid.*, 63.
42. German linguistic theory further serves to bolster our thesis of art as the communicative/existential paradigm that becomes the creative calculus of reality. For more on this theme see Nerlich and Clarke, "Mind, Meaning and Metaphor", *History of the Human Sciences* 14. 2 (2001): 39–61.
43. Bennett, *Hugo von Hofmannsthal*, 117.
44. Hofmannsthal, quoted in the introduction to Hofmannsthal, *Poems and Verse Plays*, xxv.
45. Schoenberg, *Coherence, Counterpoint, Instrumentation in Form*, 23.
46. Hofmannsthal, quoted in Bennett, *Hugo von Hofmannsthal*, 105.
47. *Ibid.*, 105–106.
48. Hofmannsthal, *The Lord Chandos and Other Writings*, 127, translation modified.
49. Altenberg, "Individualität", in Wunberg and Braakenburg, ed., *Die Wiener Moderne*, 425, my translation.
50. The turn of the century was under the spell of durational philosophy, especially as elaborated by Henri Bergson. Many such descriptions of the connection between time and space anticipate and mirror modern physics.
51. Broch, *The Guiltless*.
52. *Ibid.*, 11.
53. *Ibid.*, 26.
54. *Ibid.*, 27.
55. In his Diaries, Musil writes, "[Hermann] *Broch*: Leaning. When one leans against a wall, one's suit gets covered in white stains. Without it being plagiarism", in Musil, *Tagebücher, Aphorismen, Essays und Reden*, 361.
56. Hamburger, introduction to Hofmannsthal, *Poems and Verse Plays*, lv.
57. Broch, *The Guiltless*, 19.
58. *Ibid.*, 18–19.
59. *Ibid.*, 14.
60. *Ibid.*, 17.
61. *Ibid.*, 19.
62. *Ibid.*, 20.
63. *Ibid.*

64. *Ibid.*, 22.
65. *Ibid.*, 23.
66. Hamburger, introduction to Hofmannsthal, *Poems and Verse Plays*, xl.
67. Broch, *Guiltless*, 24.
68. Hamburger, introduction to Hofmannsthal, *Poems and Verse Plays*, xl.
69. Richard Brinkmann, "Hofmannsthal und die Sprache", quoted in Bennett, *Hugo von Hofmannsthal*, 121.

NOTES TO CHAPTER 8

1. Stuckenschmidt, *Schoenberg, His Life, World and Work*, 183.
2. Jones, *Sigmund Freud, Life and Work*, 2: 88–89.
3. See Wickes, "Schoenberg, *Erwartung*, and the Reception of Psychoanalysis in Musical Circles in Vienna until 1910/11", *Studies in Music*, 88–106.
4. Stuckenschmidt, *Schoenberg, His Life, World and Work*, 93.
5. Goehr, "Schoenberg and Karl Kraus", *Music Analysis* 4 (March-July, 1985), 64.
6. Steiner, "Schoenberg's 'Moses und Aron'", in *Language and Silence*, 155.
7. "Psychoanalyse und Musik", from the Archives of the Arnold Schoenberg Center in Vienna [ca. 1932], T39.23. Used by permission of Belmont Music Publishers. All translations are mine. Hereafter cited in text.
8. Note that the German *das Motiv*, apart from the foregoing rendition, can also mean (depending on the context) an artistic motif or a theme. As it is difficult to know, given the brevity of the phrase, which of the two meanings is preponderant, both should be kept in mind, as they suggest themselves in the original.
9. Schoenberg, *Coherence, Counterpoint, Instrumentation, Instruction in Form*, 27.
10. Schoenberg writes the following in a letter to Hugo Leichtentritt: "I am no 'reader' and therefore actually know the following books only superficially [a short list follows] and in most cases only certain sections, out of context. Nevertheless I remember many a good idea." *Letters*, 207.
11. His engineering projects span writing instruments and furniture, ending with chess and a design for a musical typewriter.
12. Schoenberg, *Style and Idea*, 269.
13. Schoenberg, "Ornaments and Construction", in *Style and Idea*, 312.
14. Dahlhaus, "Schoenberg's Aesthetic Theology", in *Schoenberg and the New Music*, 87.
15. Agamben, *The Time That Remains*, 98.
16. Schoenberg, quoted in Auner, *A Schoenberg Reader*, 76.
17. Schoenberg, *Style and Idea*, 77.
18. *Ibid.*, 78.
19. Auner ed., *A Schoenberg Reader*, 71.
20. See Stuckenschmidt, *Schoenberg, His Life, World and Work*, 21.
21. Schoenberg, *Theory of Harmony*, 21.
22. *Ibid.*, 8.
23. *Ibid.*, 2, my emphasis.
24. *Ibid.*, 116.
25. *Ibid.*, 75.
26. *Ibid.*, 53.
27. *Ibid.*, 76.
28. *Ibid.*, 77.
29. *Ibid.*, 76.

30. *Ibid.*
31. *Ibid.*, 127.
32. *Ibid.*, 128–129, my emphasis.
33. *Ibid.*, 127–128, my emphasis, translation modified.
34. *Ibid.*, 319.

NOTES TO CHAPTER 9

1. Schoenberg, *Theory of Harmony*, 403.
2. *Ibid.*, 389n.
3. See *ibid.*, 318.
4. *Ibid.*, 5.
5. *Ibid.*, 421–422, my emphasis.
6. Lefebvre, *The Production of Space*, 25.
7. José Saramago, *All the Names*, epigraph.
8. Schoenberg, *Theory of Harmony*, 392.
9. Schoenberg, *Theory of Harmony*, 352.
10. *Ibid.*, 258.
11. *Ibid.*, 318.
12. *Ibid.*, 416.
13. See *ibid.*, 416, translator's note.
14. Weber, *The Legend of Freud*, 7.
15. Schoenberg, *Theory of Harmony*, 286, my emphasis.
16. *Ibid.*, 400.
17. *Ibid.*, 313.
18. *Ibid.*, 287.
19. *Ibid.*, 319.
20. *Ibid.*, 385, my emphasis.
21. *Ibid.*, 270.
22. Agamben, *The Time That Remains*, 105.
23. Schoenberg, *Theory of Harmony*, 315.
24. Simmel, "The Problem of Historical Time", in *Essays on Interpretation in Social Science*, Guy Oaks ed., 143.
25. Schoenberg, *Theory of Harmony*, 319.
26. Schoenberg, *The Musical Idea*, 253.
27. Schoenberg, *Theory of Harmony*, 410.
28. *Ibid.*, 399.
29. *Ibid.*, 126.
30. *Ibid.*
31. The paper first appeared on the pages of *Imago* (3 [1]: 15–36) accompanied by a footnote, drafted by Freud himself: "Although this paper does not, strictly speaking, conform to the conditions under which contributions are accepted for publications in this journal, the editors have decided to print it, since the author, who is personally known to them, moves in psycho-analytic circles, and since his mode of thought has in point of fact a certain resemblance to the methodology of psycho-analysis" (SE 13: 211n).
32. Schoenberg, quoted in Crawford and Crawford, *Expressionism in Twentieth-Century Music*, 6.
33. See Steiner, *Language and Silence*, 156.
34. Schoenberg, *Theory of Harmony*, 395.
35. See Schoenberg, *Style and Idea*, 218.
36. *Ibid.*, 317.
37. Stuckenschmidt, *Schoenberg, His Life, World and Work*, 106.

38. See Schoenberg, "Composition with Twelve Tones", in *Style and Idea*, 220.
39. *Ibid.*, 223.
40. Kokoschka, quoted in Caws, ed., *Manifesto: A Century of 'Isms'*, 252.
41. Schoenberg, *Style and Idea*, 218, my emphasis.
42. Schoenberg, *Theory of Harmony*, 289–290.
43. See *ibid.*, 204.

NOTES TO THE CONCLUSION

1. Schoenberg, *Theory of Harmony*, 411–412.
2. *Ibid.*, 402.
3. *Ibid.*, 87.
4. Badiou, *The Century*, 32.
5. Badiou, *The Century*, 94.
6. Virilio, *Art and Fear*, 63.
7. Anton von Webern, quoted in Stuckenschmidt, *Schoenberg*, 84.
8. Schoenberg, *Theory of Harmony*, 412.

Bibliography

Aars, Kristian B. R. "Die Erwartung". *Zeitschrift für Psychologie and Physiologie der Sinnesorgane* 22 (1899): 401–414.

Adorno, Theodor W. "On the Fetish-Character of Music and the Regression of Listening". In *Essays on Music*, edited by Richard Leppert, 288–317. Berkeley: University of California Press, 2002.

———. *Aesthetic Theory.* Translated by Robert Hullot-Kentor. Minneapolis: University of Minnesota Press, 1999.

———. *The Philosophy of New Music.* Translated by Robert Hullot-Kentor. Minneapolis: University of Minnesota Press, 2006.

———. *Prisms.* Translated by Samuel and Shierry Weber. Cambridge, MA: MIT, 1967.

Agamben, Giorgio. *The Time That Remains: A Commentary on the Letter to the Romans.* Translated by Patricia Dailey. Stanford, CA: Stanford University Press, 2005.

Altenberg, Peter. "Individualität". In *Die Wiener Moderne: Literatur, Kunst und Musik zwischen 1890 und 1910*, edited by Gotthart Wunberg and Johannes J. Braakenburg, 424–425. Stuttgart: Reclam, 2000.

Aragon, Louis and André Breton. "Le Cinquntenaire de l'Hysterie (1878–1928)". *La Révolution Surrealiste,* 11 (1928): 20–22.

Auner, Joseph, ed. *A Schoenberg Reader: Documents of a Life.* New Haven: Yale University Press, 2003.

Badiou, Alain. *The Century.* London: Polity, 2007.

Bahr, Hermann. *Expressionismus.* Munich: Delphin-Verlag, 1920.

———. *Kulturprofil der Jahrhundertwende: Essays von Hermann Bahr.* Edited by Heinz Kindermann. Vienna: H. Bauer-Verlag, 1962.

———. *Prophet der Moderne: Tagebücher 1888–1904.* Edited by Reinhard Farkas. Vienna: Böhlau Verlag, 1987.

———. *Zur Überwindung des Naturalismus: Theoretische Schriften 1887–1904.* Edited by Gotthart Wunberg. Stuttgart: W. Kohlhammer Verlag, 1968.

Beller, Steven, ed. *Rethinking Vienna 1900.* New York: Berghahn Books, 2001.

Benjamin, Walter. *Illuminations: Essays and Reflections.* Edited by Hannah Arendt. Translated by Harry Zohn. New York: Schocken Books, 1969.

Bennett, Benjamin. *Hugo von Hofmannsthal: The Theatres of Consciousness.* Cambridge: Cambridge University Press, 1988.

Berg, Alban, et al. *Arnold Schönberg: In höchster Verehrung.* Munich: R. Piper, 1912.

Berry, Duncan J. "From Historicism to Architectural Realism: On Some of Wagner's Sources". In *Otto Wagner: Reflections on the Raiment of Modernity*, edited by Harry Francis Mallgrave, 243–278. Santa Monica, CA: Getty Center for the History of Art and the Humanities, 1993.

Bettelheim, Bruno. *Freud's Vienna & Other Essays*. New York: Vintage Books, 1989.

Blanchot, Maurice. *The Writing of the Disaster*. Translated by Ann Smock. Lincoln: University of Nebraska Press, 1995.

Blom, Philip. *The Vertigo Years: Europe 1900–1914*. New York: Basic Books, 2008.

Bouveresse, Jacques. *Wittgenstein Reads Freud: The Myth of the Unconscious*. Translated by Carol Cosman. Princeton, NJ: Princeton University Press, 1995.

Breton, André. *Nadja*. Translated by Richard Howard. New York: Grove Press, 1960.

Broch, Hermann. *The Guiltless*. Translated by Ralph Manheim. Boston: Little, Brown, 1974.

———. *The Unknown Quantity*. Translated by Willa and Edwin Muir. New York: Howard Fertig, 1975.

Brodsky Lacour, Claudia. *Lines of Thought: Discourse, Architectonics, and the Origin of Modern Philosophy*. Durham: Duke University Press, 1996.

Bujić, Bojan. *Arnold Schoenberg*. London: Phaidon, 2011.

Busoni, Ferruccio. *The Essence of Music and Other Papers*. Translated by Rosamond Ley. New York: Dover, 1987.

Calinescu, Matei. *Five Faces of Modernity: Modernism, Avant-Garde, Decadence, Kitsch, Postmodernism*. Durham: Duke University Press, 1987.

Cavell, Richard. *McLuhan in Space: A Cultural Geography*. Toronto: University of Toronto Press, 2002.

Caws, Mary Ann, ed. *Manifesto: A Century of Isms*. Lincoln: University of Nebraska Press, 2001.

Christensen, Thomas. *Rameau and Musical Thought in the Enlightenment*. Cambridge: Cambridge University Press, 1993.

Colomina, Beatriz. *Privacy and Publicity: Modern Architecture as Mass Media*. Cambridge, MA: MIT, 1996.

Crary, Jonathan. *Suspension of Perception: Attention, Spectacle, and Modern Culture*. Cambridge, MA: MIT, 2001.

Crawford, John C. "Die Glückliche Hand: Further Notes". *Journal of the Arnold Schoenberg Institute* 4.1 (1980): 69–76.

Crawford, John C., and Dorothy L. Crawford. *Expressionism in Twentieth-Century Music*. Bloomington: Indiana University Press, 1993.

Dahlhaus, Carl. *Schoenberg and the New Music*. Translated by Derrick Puffett and Alfred Clayton. Cambridge: Cambridge University Press, 1988.

Damisch, Hubert. *The Origin of Perspective*. Translated by John Goodman. Cambridge, MA: MIT Press, 2000.

Deleuze, Gilles, and Félix Guattari. *A Thousand Plateaus: Capitalism and Schizophrenia*. Minneapolis: University of Minnesota Press, 1987.

———. *What Is Philosophy?* New York: Columbia University Press, 1994.

Derrida, Jacques. "Force and Signification". In *Writing and Difference*, translated by Alan Bass, 3–30. Chicago: University of Chicago Press, 1978.

Descartes, René. *Meditations on First Philosophy*. Translated by Donald A. Cress. Indianapolis: Hackett, 1993.

Didi-Huberman, Georges. *Invention of Hysteria: Charcot and the Photographic Iconography of the Salpêtrière*. Translation by Alisa Hartz. Cambridge, MA: MIT, 2003.

Downes, Stephen. *Music and Decadence in European Modernism: The Case of Central and Eastern Europe*. Cambridge: Cambridge University Press, 2010.

Düchting, Hajo. *Paul Klee: Malerei und Musik*. Munich: Prestel, 2001.

Endell, August. "Möglichkeiten und Ziele einer neuen Architektur". *Deutsche Kunst und Dekoration* 1 (1898): 141–152.

Erlmann,Veit. *Reason and Resonance: A History of Modern Aurality*. New York: Zone Books, 2010.
Focillon, Henri. *The Life of Forms in Art*. 1934. Translated by Charles Beecher Hogan and George Kubler. New York: Zone Books, 1992.
Foucault, Michel. *The Archaeology of Knowledge*. Translated by A. M. Sheridan-Smith. London: Routledge, 1989.
———. *Discipline and Punish*. Translated by Alan Sheridan. New York: Vintage Books, 1991.
———. *Madness and Civilization*. Translated by Richard Howard. New York: Vintage, 1988.
———. *The Order of Things: An Archaeology of the Human Sciences*. London: Routledge, 2002.
Freud, Sigmund. *The Standard Edition of the Complete Psychological Works of Sigmund Freud*. Translated and edited by James Strachey, in collaboration with Anna Freud, with the assistance of Alix Strachey and Alan Tyson. 24 vols. London: Hogarth, 1955.
Frisby, David. *Fragments of Modernity: Theories of Modernity in the Work of Simmel, Kracauer and Benjamin*. Cambridge, MA: MIT Press, 1988.
Frisch, Walter. "Music and Jugendstil". *Critical Inquiry* 17 (1990): 138–161.
Gafijczuk, Dariusz. "Bending Modernity: Psychoanalysis and the Rest of Culture". *Journal of Historical Sociology* 22. 4 (2009): 447–475.
———. "Dwelling Within: The Inhabited Ruins of History". *History & Theory* 52.2 (May 2013): 149–170.
Gay, Peter. *Freud: A Life for Our Time*. New York: W. W. Norton, 1998.
———. *Schnitzler's Century: The Making of Middle-Class Culture 1815–1914*. London: Allen Lane, 2001.
Goehr, Alexander. "Schoenberg and Karl Kraus". *Music Analysis* 4 (March–July 1985): 59–71.
Hamburger, Michael, ed. *Hugo von Hofmannsthal—Poems and Verse Plays*. London: Routledge & Kegan Paul, 1961.
Hanslick, Eduard. *The Beautiful in Music*. 1854. Translated by Gustav Cohen. Indianapolis: Bobbs-Merrill, 1957.
Harrison, Thomas. *1910—The Emancipation of Dissonance*. Berkeley, CA: University of California Press, 1996.
Helmholtz, Hermann L. F. *Helmholtz on Perception*. Edited by Richard Warren and Roslyn Warren. New York: Wiley & Sons, 1968.
———. *On the Sensations of Tone as a Physiological Basis for the Theory of Music*. 1863. Translated by Alexander J. Ellis. New York: Dover, 1954.
Henry, Michel. *The Genealogy of Psychoanalysis*. Translated by Douglas Brick. Stanford, CA: Stanford University Press, 1993.
Hermand, Jost, ed. *Jugendstil*. Darmstadt, Germany: Wissenschaftliche Buchgesellschaft, 1971.
Hofmannsthal, Hugo von. *Poems and Verse Plays*. Edited by Michael Hamburger. London: Routledge & Kegan Paul, 1961.
Huter, Michael. "Body as Metaphor: Aspects of the Critique and Crisis of Language at the Turn of the Century with Reference to Egon Schiele". In *Egon Schiele: Art, Sexuality, and Viennese Modernism*, edited by Patrick Werkner, 119–129. Palo Alto, CA: The Society for the Promotion of Science and Scholarship, 1994.
———. *The Lord Chandos and Other Writings*. Translated by Joel Rotenberg and John Banville. New York: New York Review of Books, 2005.
Janik, Allan, and Stephen Toulmin. *Wittgenstein's Vienna*. New York: Simon and Schuster, 1973.
Jankélévitch, Vladimir. *Music and the Ineffable*. Translated by Carolyn Abbate. Princeton, NJ: Princeton University Press, 2003.

Jay, Martin. *Downcast Eyes: The Denigration of Vision in Twentieth-Century French Thought*. Berkeley, CA: University of California Press, 1994.
Jones, Ernest. *Sigmund Freud: Life and Work*. 3 vols. London: Hogarth, 1953.
Kahn, Douglas. *Noise. Water. Meat. A History of Sound in the Arts*. Cambridge, MA: MIT, 2001.
Kandinsky, Wassily. *Point and Line to Plane*. New York: Dover, 1979.
Kandinsky, Wassily, and Franz Marc, eds. *Der blaue Reiter*. 1912. Munich: Piper, 2004.
Kierkegaard, Soren. *Fear and Trembling and the Sickness unto Death*. Translated by Walter Lowrie. Princeton, NJ: Princeton University Press, 1969.
Kittler, Friedrich A. *Discourse Networks 1800/1900*. Translated by Chris Cullen and Michael Metteer. Stanford, CA: Stanford University Press, 1992.
Kraus, Karl. *Dicta and Contradicta*. Translated by Jonathan McVity. Urbana: University of Illinois Press, 2001.
———. *Die letzten Tage der Menschheit*. 1921. Frankfurt, Germany: Suhrkamp, 1986.
Kundera, Milan. *The Art of the Novel*. London: Faber and Faber, 1988.
Kwinter, Sanford. *Architectures of Time: Toward a Theory of the Event in Modernist Culture*. Cambridge, MA: MIT, 2002.
Laborda, José Maria Garcia. *Studien zu Schönbergs Monodram "Erwartung" op. 17*. Laaber: Laaber-Verlag, 1981.
Lacoue-Labarthe, Philippe. "The Echo of the Subject". In *Typography: Mimesis, Philosophy, Politics*, translated by Christopher Fynsk, 139–207. Stanford, CA: Stanford University Press, 1998.
Lange, Konrad. "Gedanken zu einer Ästhetick auf entwickelungsgeschichtlicher Grundlage. Gleichzeitig als Bericht über Karl Gross, 'Die Spiele der Tiere'". *Zeitschrift für Psychologie and Physiologie der Sinnesorgane* [Journal for the Psychology and Physiology of the Sense Organs] 14 (1897): 242–273.
Laplanche, J., and J. B. Pontalis. *Das Vocabular der Psychoanalyse*. Translated by Emma Moersch. Frankfurt, Germany: Suhrkamp, 1973.
Lefebvre, Henri. *The Production of Space*. Translated by Donald Nicholson-Smith. Oxford: Blackwell, 1991.
Lipps, Theodor. "Aesthetische Einfühlung". *Zeitschrift für Psychologie and Physiologie der Sinnesorgane* [Journal for the Psychology and Physiology of the Sense Organs] 22 (1900): 415–450.
———. *Consonance and Dissonance in Music*. Translated by William Thomson. San Marino, CA: Everett Books, 1995.
Loos, Adolf. *Ins Leere gesprochen: Gesammelte Schriften 1897–1900*. 1921. Edited by Adolf Opel. Vienna: Georg Prachner Verlag, 1997.
———. *Ornament und Verbrechen: Ausgewählte Schriften*. Edited by Adolf Opel. Vienna: Georg Prachner Verlag, 2000.
———. *Trotzdem: Gesammelte Schriften 1900–1930*. 1931. Edited by Adolf Opel. Vienna: Georg Prachner Verlag, 1997.
Luft, David S. *Eros and Inwardness in Vienna: Weininger, Musil, Doderer*. Chicago: University of Chicago Press, 2003.
Lukács, György. *Goethe and His Age*. Translated by R. Anchor. London: Merlin Press, 1968.
———. *Soul and Form*. Translated by John T. Sanders and Katie Terezakis. New York: Columbia University Press, 2010.
Lütkehaus, Ludger, ed. *Dieses wahre innere Afrika: Texte zur Entdeckung des Unbewussten vor Freud*. Giessen, Germany: Imago, Psychosozial-Verlag, 2005.
Mach, Ernst. *The Analysis of Sensations, and the Relation of the Physical to the Psychical*. Chicago: Open Court, 1897.

———. "Antimetaphysische Vorbemerkungen". In *Die Wiener Moderne: Literatur, Kunst und Musik zwischen 1890 und 1910*, edited by Gotthart Wunberg and Johannes J. Braakenburg, 137–145. Stuttgart: Reclam, 2000.

Mallgrave, Harry Francis, ed. *Otto Wagner: Reflections on the Raiment of Modernity*. Santa Monica, CA: Getty Center for the History of Art and the Humanities, 1993.

Merleau-Ponty, Maurice. *Visible and the Invisible*. Translated by Alphonso Lingis. Evanston: Northwestern University Press, 1968.

Moravánszky, Ákos. "Aesthetics of the Mask". In *Otto Wagner: Reflections on the Raiment of Modernity*, edited by Harry Francis Mallgrave, 199–239. Santa Monica, CA: Getty Center for the History of Art and the Humanities, 1993.

Musil, Robert. *Diaries 1899–1941*. Translated by Philip Payne. New York: Basic Books, 1998.

———. *The Man without Qualities*. Translated by Eithne Wilkins and Ernst Kaiser. 3 vols. London: Pan Books, 1987.

———. *Tagebücher, Aphorismen, Essays und Reden*. Edited by Adolf Frisé. Hamburg: Rowohlt Verlag, 1955.

Nancy, Jean-Luc. *Listening*. Translated by Charlotte Mandell. New York: Fordham University Press, 2007.

Nerlich, Brigitte, and David D. Clarke. "Mind, Meaning and Metaphor: The Philosophy and Psychology of Metaphor in 19th-Century Germany". *History of the Human Sciences* 14.2 (2001): 39–61.

Neumeyer, Fritz. "Iron and Stone: The Architecture of the *Großstadt*". In *Otto Wagner: Reflections on the Raiment of Modernity*, edited by Harry Francis Mallgrave, 115–153. Santa Monica, CA: Getty Center for the History of Art and the Humanities, 1993.

Nietzsche, Friedrich. "On the Uses and Disadvantages of History for Life". In *Untimely Meditations*, translated by R.J. Hollingdale, 59–123. Cambridge: Cambridge University Press, 1997.

———. *The Birth of Tragedy and the Case of Wagner*. Translated by Walter Kaufmann. New York: Vintage Books, 1967.

———. *Human All Too Human: A Book for Free Spirits*. Translated by R. J. Hollingdale. Cambridge: Cambridge University Press, 1996.

———. *On the Genealogy of Morals and Ecce Homo*. Translated by Walter Kaufman. New York: Vintage Books, 1989.

———. *Thus Spoke Zarathustra: A Book for None and All*. Translated by Walter Kaufmann. London: Penguin Books, 1978.

Noudelmann, François. *The Philosopher's Touch: Sartre, Nietzsche and Barthes at the Piano*. Translated by Brian J. Reilly. New York: Columbia University Press, 2012.

Panofsky, Erwin. *Perspective as Symbolic Form*. 1927. Translated by Christopher S. Wood. New York: Zone Books, 1997.

Rancière, Jacques. *The Politics of Aesthetics*. Translated by Gabriel Rockhill. London: Continuum, 2006.

Ricoeur, Paul. *Freud and Philosophy: An Essay on Interpretation*. Translated by Denis Savage. New Haven: Yale University Press, 1970.

Riegl, Alois. *Historical Grammar of the Visual Arts*. Translated by Jacqueline E. Jung. New York: Zone Books, 2004.

———. *Problems of Style: Foundations for a History of Ornament*. Translated by Evelyn Kain. Princeton, NJ: Princeton University Press, 1992.

———. *Spätrömische Kunstindustrie*. Darmstadt: Wissenschaftliche Buchgesellschaft, 1973.

Rilke, Rainer M. *Letters to a Young Poet*. Translated by Stephen Mitchell. New York: Vintage Books, 1986.

Rognoni, Luigi. *The Second Vienna School: Expressionism and Dodecaphony.* Translated by Robert W. Mann. London: John Calder, 1977.

Ruskin, John. *The Seven Lamps of Architecture.* London: Smith, Elder, 1849.

Russell, Bertrand. *A History of Western Philosophy.* New York: Simon & Schuster, 1972.

Saramago, José. *All the Names.* Translated by Margaret Jull Costa. New York: Harcourt, 1999.

Sayer, Derek. *The Coasts of Bohemia: A Czech History.* Princeton, NJ: Princeton University Press, 1998.

———. *Prague, the Capital of the Twentieth Century: A Surrealist History.* Princeton, NJ: Princeton University Press, 2013.

Schnitzler, Arthur. *Night Games and Other Stories and Novellas.* Translated by Margret Schaefer. Chicago: Ivan R. Dee, 2002.

Schoenberg, Arnold. *Coherence, Counterpoint, Instrumentation in Form.* Edited by Severine Neff. Lincoln: University of Nebraska Press, 1994.

———. *Erwartung: Monodram in einem Akt.* Vienna: Universal Edition, 1909.

———. *Letters.* Edited by Erwin Stein. Translated by Eithne Wilkins and Ernst Kaiser. London: Faber, 1974.

———. *The Musical Idea and the Logic, Technique, and Art of Its Presentation.* Edited and translated by Patricia Carpenter and Severine Neff. New York: Columbia University Press, 1995.

———. "Psychoanalyse und Musik" [ca. 1932], T39.23, Archives of the Arnold Schoenberg Center, Vienna.

———. *Style and Idea: Selected Writings of Arnold Schoenberg.* Edited by Leonard Stein. Translated by Leo Black. Berkeley: University of California Press, 1975.

———. *Theory of Harmony.* 1911. Translated by Roy E. Carter. Berkeley, CA: University of California Press, 1983.

Schorske, Carl E. *Fin-de-Siècle Vienna: Politics and Culture.* New York: Vintage Books, 1981.

Serres, Michel. *The Parasite.* Translated by Lawrence R. Schehr. Minneapolis: University of Minnesota Press, 2007.

Shawn, Arnold. *Arnold Schoenberg's Journey.* Cambridge, MA: Harvard University Press, 2002.

Simmel, Georg. "The Conflict in Modern Culture". In *The Conflict in Modern Culture and Other Essays*, translated by K. Peter Etzkorn, 11–25. New York: Teachers College Press, 1968.

———. "The Ruin". In *Essays on Sociology, Philosophy and Aesthetics*, edited by Kurt H. Wolff, 259 -266. New York: Harper & Row, 1965.

———. "Sociological Aesthetics". In *The Conflict in Modern Culture and Other Essays*, translated by K. Peter Etzkorn, 68–85. New York: Teachers College Press, 1968.

———. *Philosophische Kultur: Gesammelte Essais.* Leipzig: Verlag von Dr. Werner Klinkhardt, 1911.

———. "Das Problem des Stiles". *Die Kunst* 19 (1908): 307–316.

———. "The Problem of Historical Time". In *Essays on Interpretation in Social Science*, edited and translated by Guy Oakes, 127–144. Manchester: Manchester University Press, 1980.

———. *Simmel on Culture.* Edited by David Frisby and Mike Featherstone. London: SAGE, 1997.

Simms, Bryan R. *The Atonal Music of Arnold Schoenberg 1908–1923.* Oxford: Oxford University Press, 2000.

———. "Whose Idea Was *Erwartung*?" In *Constructive Dissonance: Arnold Schoenberg and the Transformations of Twentieth-Century Culture*, edited by

Juliane Brand and Christopher Hailey, 100–111. Berkeley, CA: University of California Press, 1997.

Smith, Gary, ed. *On Walter Benjamin: Critical Essays and Recollections*. Cambridge, MA: MIT, 1991.

Steiner, George. "Schoenberg's Moses und Aron". In *Language and Silence*, 150–163. London: Faber and Faber, 1967.

Sternberger, Dolf. "Jugendstil—Begriff und Physiognomik". In *Jugendstil*, edited by Jost Hermand, 27–46. Darmstadt: Wissenschaftliche Buchgesellschaft, 1971.

———. *Ueber den Jugendstil und andere Essays*. Hamburg: Claassen Verlag, 1956.

Strindberg, August. *Selected Essays*. Edited and translated by Michael Robinson. Cambridge: Cambridge University Press, 1996.

Stuckenschmidt, H. H. *Schoenberg, His Life, World and Work*. Translated by Humphrey Searle. London: John Calder, 1977.

Sussman, Henry. *The Aesthetic Contract: Status of Art and Intellectual Work in Modernity*. Stanford. CA: Stanford University Press, 1997.

Szendy, Peter. *Listen: A History of Our Ears*. Translated by Charlotte Mandell. New York: Fordham University Press, 2008.

Tarasti, Eero. "Metaphors of Nature and Organicism in the Epistemology of Music: A 'biosemiotic' Introduction to the Analysis of Jean Sibelius' Symphonic Thought. *Sign System Studies*, 29.2 (2001): 657–682.

Tichy, Marina, and Sylvia Zwettler-Otte. *Freud in der Presse: Rezeption Sigmund Freuds und der Psychoanalyse in Oesterreich 1895–1938*. Vienna: Sonderzahl, 1999.

Uexküll, Jakob von. *Umwelt und Innenwelt der Tiere*. Berlin: Verlag von Julius Springer, 1909.

———. "An Introduction to Umwelt". *Semiotica* 134–1/4 (2001): 107–110.

Virilio, Paul. *The Aesthetics of Disappearance*. Translated by Philip Beitchman. Cambridge. MA: Semiotext(e), 1991.

———. *Art and Fear*. Translated by Julie Rose. London: Continuum, 2003.

Wagner, Otto. *Modern Architecture: A Guide Book for His Students to This Field of Art*. 1902. Translated by Harry Francis Mallgrave. Chicago: University of Chicago Press, 1988.

Weber, Samuel. *The Legend of Freud*. Stanford, CA: Stanford University Press, 2000.

Wellesz, Egon. *Arnold Schoenberg: The Formative Years*. London: Greenwood Press, 1925.

Werkner, Patrick, ed. *Egon Schiele: Art, Sexuality, and Viennese Modernism*. Palo Alto, CA: Society for the Promotion of Science and Scholarship, 1994.

Whyte, Lancelot Law. *The Unconscious before Freud*. New York: Anchor Books, 1962.

Wickes, Lewis. "Schoenberg, *Erwartung*, and the Reception of Psychoanalysis in Musical Circles in Vienna until 1910/11". *Studies in Music* 23 (1989): 88–106.

Wistrich, Robert S. *Laboratory for World Destruction: Germans and Jews in Central Europe*. Lincoln: University of Nebraska Press, 2007.

Wittgenstein, Ludwig. *Wittgenstein's Lectures, Cambridge 1932–1935: From the Notes of Alice Ambrose and Margaret MacDonald*. Edited by Alice Ambrose. Oxford: Blackwell, 1979.

Wood, Christopher S., ed. *Vienna School Reader: Politics and Art Historical Method in the 1930s*. New York: Zone Books, 2003.

Worringer, Wilhelm. *Abstraction and Empathy: A Contribution to the Psychology of Style*. 1905. Translated by Michael Bullock. New York: International Universities Press, 1953.

Wunberg, Gotthart, and Johannes J. Braakenburg, eds. *Die Wiener Moderne: Literatur, Kunst und Musik zwischen 1890 und 1910*. Stuttgart, Germany: Reclam, 2000.

Zola, Émile. *The Experimental Novel and Other Essays*. 1880. Translated by Belle M. Sherman. New York: Haskell House, 1964.

Index